Philosophical Reflections on Medical Ethics

Philosophical Reflections on Medical Ethics

Edited by

Nafsika Athanassoulis
Lecturer in Ethics, Centre for Professional Ethics, Keele University

First published in hardback 2005
This paperback edition published 2010 by
PALGRAVE MACMILLAN

Palgrave Macmillan in the UK is an imprint of Macmillan Publishers Limited, registered in England, company number 785998, of Houndmills, Basingstoke, Hampshire RG21 6XS.

Palgrave Macmillan in the US is a division of St Martin's Press LLC, 175 Fifth Avenue, New York, NY 10010.

Palgrave Macmillan is the global academic imprint of the above companies and has companies and representatives throughout the world.

Palgrave® and Macmillan® are registered trademarks in the United States, the United Kingdom, Europe and other countries.

ISBN 978-0-230-24704-8   ISBN 978-0-230-27393-1 (eBook)
DOI 10.1007/978-0-230-27393-1

This book is printed on paper suitable for recycling and made from fully managed and sustained forest sources. Logging, pulping and manufacturing processes are expected to conform to the environmental regulations of the country of origin.

A catalogue record for this book is available from the British Library.

Library of Congress Cataloging-in-Publication Data
Philosophical reflections on medical ethics / edited by Nafsika Athanassoulis.
    p. cm.
  Includes bibliographical references and index.
  ISBN 978-1-4039-4527-3 (cloth)
    1. Medical ethics. 2. Medical ethics – Philosophy. I.
  Athanassoulis, Nafsika, 1973–

R725.5.P48 2005
174.2–dc22   2005043760

10   9   8   7   6   5   4   3   2   1
19   18   17   16   15   14   13   12   11   10
Transferred to Digital Printing 2010

Contents

Acknowledgements

This volume is the result of four years of thinking about and teaching medical ethics, so my greatest debt is owed to the hundreds of students at the University of Leeds who have stimulated my own research with their arguments and ideas. I am also particularly grateful to all the contributors to this volume for their hard work and excellent papers.

Notes on the Contributors

Nafsika Athanassoulis is Lecturer in Ethics at Keele University. She is the author of *Morality, Moral Luck and Responsibility: Fortune's Web* (Palgrave Macmillan, 2005), and has published articles on moral luck, virtue ethics and consent. She teaches medical ethics to anyone who will listen, including philosophy students, medical students, biomedical sciences students and professionals returning to education.

Piers Benn is Lecturer in Medical Ethics and Law, Imperial College. He is the course organiser for the new MSc in Medical Ethics and has general interests in moral theory, meta-ethics and applied ethics. His research interests in medical ethics include issues at the beginning and end of life, including the metaphysics of 'personhood' and the debate about brain death; ethical issues in psychiatry, especially the moral responsibility of psychopaths; the nature and defensibility of paternalism in medicine; lying and deception in medicine and the interplay of secular and theological approaches to biological concerns. His publications include *Ethics*, (1998); 'Morality, the Unborn and the Open Future', in *Questions of Time and Tense*, edited by R. LePoidevin (1998); 'New Issues: Health Care Ethics', *Journal of Applied Philosophy* 18:2 (2001); 'Medicine, Lies and Deceptions', *Journal of Medical Ethics* 27:2 (2001). He is currently working on a book on death.

Katrien Devolder is a postdoctoral research fellow of the Research Foundation – Flanders and is based at the Bioethics Institute Ghent at Ghent University, Belgium. Her chief interests are in bioethics, more particularly in cloning, stem cell research and genetics. She is the author, together with Johan Braeckman, of a book on human cloning, which was published in 2001. She is currently writing a book on the ethics of human embryonic stem cell research and has published extensively on this topic.

Heather Draper is Reader in Biomedical Ethics, Centre for Biomedical Ethics, Primary Care Clinical Sciences, at the University of Birmingham, UK. One of her primary research interests is the ethics of human reproduction and the role in ethics and policy of genetic relatedness in determining parental rights and responsibilities. Published papers related to the one in this collection include: 'Paternity Fraud and Compensation for Misattributed Paternity' *Journal of Medical Ethics* (2007), 'Gametes, Consent and the Point of No Return' *Human Fertility* (2007), and with Jonathan Ives 'Testing for Fatherhood: Two Paradoxes of Paternity Testing' *Journal of Social Welfare and Family Law* (forthcoming) and 'Becoming a Father/Refusing Fatherhood: An Empirical Bioethics Approach to Paternal Responsibilities and Rights' *Clinical Ethics* (2008).

Ray Frey is Professor of Philosophy, Bowling Green University. He is the author of numerous articles and books on ethical theory, applied ethics, the history of ethics and social and political theory, including *Euthanasia and Physician-Assisted Suicide* (1998, with Gerald Dworkin and Sissela Bok), Awaiting publication are books on Joseph Butler, an edition of Butler's ethical writings, a volume of essays on topics in applied ethics and a volume on utilitarianism. He teaches courses in all the areas of his research, including legal philosophy. Previously, he tutored students at Oxford and taught in the Universities of Liverpool and Toronto. He is Senior Research Fellow at the Social Philosophy and Policy Center in Bowling Green, a Senior Research Fellow of the Kennedy Institute of Ethics, Georgetown University, and a Senior Research Fellow of the Westminster Institute of Ethics and Public Policy at the University of Western Ontario, Canada.

John Harris is Sir David Alliance Professor of Bioethics, School of Law, University of Manchester and is joint Editor-in-Chief of *The Journal of Medical Ethics*. He was elected a Fellow of the Academy of Medical Sciences (FMedSci) in 2001 and has been a member of the Human Genetics Commission since its foundation in 1999. Recent books include: *Clones Genes and Immortality* (1998); *Bioethics* (2001); *A Companion to Genethics: Philosophy and the Genetic Revolution* (2002 and 2004); and *On Cloning* (2004).

Hallvard Lillehammer is Fellow of King's College and Lecturer in Philosophy at Cambridge University. He is co-editor of *Real Metaphysics* (2003) and *Genetics, Persons, and Responsibility* (2001). He has published a number of papers on medical ethics, including 'Voluntary Euthanasia and the Logical Slippery Slope Argument', *The Cambridge Law Journal* (2002), and 'Who Needs Bioethicists?', *Studies in the History and Philosophy of Biological and Biomedical Sciences* (2004).

David S. Oderberg is Professor of Philosophy at the University of Reading. He is the author of many articles in metaphysics, ethics, philosophical logic and other subjects. His publications include *Moral Theory: A Non-Consequentialist Approach* (2000); *Applied Ethics: A Non-Consequentialist Approach* (2000); and *Human Lives: Critical Essays on Consequentialist Bioethics* (co-editor with Jacqueline A. Laing, 1997).

James Stacey Taylor is an Associate Professor of Philosophy at The College of New Jersey, USA. He is the editor of *Personal Autonomy: New Essays* (2005, paperback 2008), and the author of both *Stakes and Kidneys: Why Markets in Human Body Parts are Morally Imperative* (2005) and *Practical Autonomy and Bioethics* (2009). He is also the author of over 30 articles in academic journals, including *Social Philosophy & Policy, Public Affairs Quarterly, American Philosophical Quarterly*, the *Journal of Applied Ethics*, and the *Journal of Medicine and Philosophy*. His work has also appeared in many edited collections and has been widely anthologized. He currently serves as the Managing Editor of the *Journal of Value Inquiry*, and is completing two books: *Death Unterrible: Epicurean Thanatology and Contemporary Bioethics* and *Toxic Trade: An Unapologetic Defense of Universal Commodification*.

Stephen Wilkinson is Senior Lecturer in Ethics and Philosophy at Keele University, where he directs a number of graduate programmes including the PgDip/MA in Medical Ethics and Law, taught jointly with Keele University's Law Department, and the UK's first professional doctorate in Medical Ethics (the DMedEth), which was launched in 2002. His recent research papers have addressed topics including the allocation of health service resources, pre-implantation genetic diagnosis, separating conjoined twins, and the nature of mental illness. A paper on the latter won the Philosophical Quarterly International Essay Prize in 1999. He has recently completed a book on the ethics of commercialising the human body (*Bodies for Sale*, 2003). This work was supported by a research grant by the Arts and Humanities Research Board.

Introduction

Nafsika Athanassoulis

The title of this collection is deliberately misleading. Medical ethics is, quite uncontroversially, a sub-discipline of moral philosophy, and it should, therefore, be impossible to reflect on medical ethics in anything but a philosophical manner. However, this title was selected specifically to make a point about the overall approach and style of the contributions. To see this point we need to consider the recent trends in the development of medical ethics research.

If moral philosophy concerns itself with normativity, questions about moral obligation, what is the right thing to do or what kind of person I should try to become, then medical ethics is a sub-discipline of moral philosophy which asks these kinds of questions within the context of health care. The last couple of decades have seen a huge increase in interest in medical ethics research. A number of factors have contributed to this, including a broader understanding of what makes up a good curriculum in the health care sciences, changing views about the role of health care professionals and their relationship to their patients and public, changing public perceptions about the health care professions and evolving standards of accountability. This increase in interest in medical ethics and the opening up of the subject itself to attract students and researchers across disciplines have had a significant impact on academic philosophy itself. Academic philosophers, whether we like it or not, have expanded their sphere of influence, teaching across disciplines, engaging in interdisciplinary work and bringing their research, for the first time in a rather long while, out of the rather insular space of academia and into a more public domain.

On the whole this has been an extremely positive influence on academic philosophy, forcing us to engage with current topics, become

1

involved in public debates and have productive contact with colleagues in other disciplines. The first step in this process of engaging philosophically with a wider public through the issues raised in medical ethics, has been a long process of sensitisation. Becoming aware of moral issues and the questions they raise is not an easy task. It seems to me that this first step, of recognising that there is a question to be asked and recognising the nature of this question, is by far the most demanding step in this process. The general interest in medical ethics gave philosophers the opportunity to 'awaken' a whole new audience to the kinds of questions morality raises. This sensitisation to the normative aspects of the world is a long and challenging process and requires a veritable shift in the way we perceive ourselves, the world and our relationship to it.

It seems, then, that the task of much of the research in medical ethics so far has been to raise questions, to encourage the reader to interpret situations in a particular way, i.e. in a morally relevant way, to provoke interest and concern. As a first step, this is particularly important, especially if we see it in the context of introducing a discipline to a new audience. However, it is just a first step and we should be careful that it does not come to dominate the entire tenor of the debate in medical ethics.

Raising concerns, pointing out potentially problematic areas, sensitising one's audience, is just a first step and should not take over the entire process. The next step must be a concerted effort to give answers to the problems raised. These answers are, however, difficult to arrive at and require dedication of purpose and a genuine commitment to philosophical enquiry. One of the more unfortunate developments arising from this increased interest in medical ethics is that 'ethics' has become a byword for a rather more approachable, intellectually lighter and more easily comprehensible approach to practical morality which would be acceptable to many who might be resistant to thinking philosophically as such. Thus when trying to persuade colleagues in other disciplines of the value of incorporating philosophy in their curricula, the term 'ethics' is used instead, encouraging the belief that by studying ethics one is avoiding studying the rather obscure, practically inconclusive and overall unsatisfying discipline of philosophy. This seems to me to be a mistake.

If philosophy is about reasoning, then any being that has even had a thought is, in a sense, a philosopher. Of course, philosophy is also about reasoning well, i.e. reasoning consistently, validly, persuasively, etc., so simply qualifying as a philosopher is not the end of the story.

One has to work hard to be able to make any claim to being the kind of person who can reason well. The answers are difficult to come by because of the nature of the discipline which concerns itself with the most fundamental questions about human existence. We should not, therefore, expect easy answers, but at the same time we should not shy away from the effort to find these answers. Medical ethics has now gained a wider audience, an audience which has had time to acquaint itself with the demands of philosophy as a discipline. Although becoming sensitive to the kinds of questions moral philosophy asks is an important step, it is still only a first step and more needs to be done. This is the time to introduce a wider audience to the difficulties of reasoning about the answers and we should not underestimate the philosophical sophistication of this new, wider audience. Collections on specialist areas in medical ethics, for example on issues in gynaecological medical ethics, or issues in palliative care, have their place in research. They often fulfil the role of helping their audiences become aware of philosophical issues, however we should not over-emphasise this role.

The aim of this collection is to bring together essays on medical ethics focusing on their philosophical merits, rather than on a narrow topic of interest. Perceiving one's research narrowly focused in this way, as being research only of relevance to palliative care medical ethics, is of dubious philosophical merit. If philosophy is about reasoning, then one can reason about anything. After all how can one claim to reason philosophically *only* about medical ethics and not about any other aspect of the world? No philosophical argument applies exclusively to medical ethics and anyone interested in medical ethics should embrace an interest in philosophy in general.

The contributors in this volume see themselves primarily as philosophers, as thinkers with an interest in researching good arguments. Good arguments can be of use to all areas of philosophical concern and in this collection they come to bear on questions in medical ethics.

Making decisions about how we should behave when our actions affect the wellbeing of other people is difficult enough in the more standard sort of cases. Some of these decisions are even more difficult as they have to be made in the absence of any identifiable person who will be benefited or harmed by our decisions. Decisions we make now may affect the quality of life of people who do not yet exist as well as the very identity of these individuals. For example, decisions about polluting the environment now will affect future generations. This is the non-identity problem, most famously associated with the work of Derek Parfit. Advances in reproductive technologies relating to pre-conception choices may involve making decisions about the welfare of future persons in the absence of any distinct or identifiable person. How we should make such decisions and especially whether pre-conception scenarios are similar to other non-identity scenarios is the topic of the first chapter, by Hallvard Lillehammer.

According to the person-involving principle, when benefit is produced there is an identifiable individual who benefits, and conversely in the absence of an identifiable individual no benefit can be produced. For followers of this view, pre-conception decisions are fairly unproblematic as they do not affect an identifiable individual so there is no sense in which they benefit or fail to benefit anyone. However, even if, in the absence of an identifiable individual, we cannot coherently discuss whether conception would be a benefit to this non-existent person, we can coherently evaluate the merits of different lives. One way of doing this is to suggest impartial, non-person-specific considerations about what makes a life better or worse, so that more benefit is better than less. However, this leads to the repugnant conclusion, i.e. for any world containing a number of people with good quality of life, there is another world containing more people with a less good quality of life, leading to a world with a vast number of people with very low quality of life.

Lillehammer considers and rejects various proposed solutions to the non-identity problem and argues that it is a mistake to think that we have to choose either the repugnant conclusions of the beneficence principle or seeing our choices as involving either person-involving principles or impartial non-person involving principles. What we should do is recognise that just as partial considerations apply when we make decisions involving identifiable individuals, they apply when we make decisions involving non-identifiable individuals. For pre-conception scenarios inevitably involve a partial evaluative perspective due to their particular nature. To show why this is the case Lillehammer considers a number of examples involving creating something or someone where there is a choice of having created something/someone better and uses them to discuss judgements of value which underpin the idea that one option is better than another. From an impartial perspective it is wrong to create a suboptimal tool as the tool exists for a purpose and failing to satisfy that purpose means that it fails as a tool. By comparison though, gifts are evaluated differently, within a context of gift-giving, and not purely instrumentally as the objects that they are. Similarly, evaluating the creation of a pet has to be done against a background of the environment the pet has to live in and it cannot reasonably be demanded that individuals evaluate the relevant questions, i.e. creating the pet and evaluating plural entities, such as marriages, on *purely* impartial grounds. If we bring these considerations to bear on pre-conception decisions we can see that evaluating decisions to avoid or seek certain characteristics purely on impartial grounds will not work. On impartial grounds it is clear that disabled children, for example, should not be conceived provided we accept that it is not as good to lead a disabled life as an able-bodied one. However, such decisions are not made on impartial grounds; they are made within specific historical and social contexts and should be understood as such. Claiming that one's children would be as well off being deaf in a community which offers emotional, physical and financial support to the deaf is a claim based on specific partial grounds.

Finally, Lillehammer turns his attention to the definition of disability. In order to evaluate specific decisions avoiding or selecting for disabilities, we need to look at the decision in the context in which it is made. We need to evaluate the partial reasons given for the choice, because partial values are part of non-identity decisions. Partial values not only form part of non-identity scenarios, but they should do so, as making such decisions on impartial grounds can be dangerous – dangerous because, by emphasising impartial considerations, we lose sight of the partial values which truly underlie and justify (or not) these decisions.

The next chapter, by Stephen Wilkinson, adopts a different approach on a similar topic. Whereas Lillehammer focuses on a specific argument relevant to particular kinds of decision, Wilkinson's broad argumentative strategy is a reduction. He examines the arguments mainly against, but also some arguments for, so-called 'designer babies' by considering an impressive number of alternatives. Some objections considered collapse into larger ones which do not apply exclusively to designer babies and are therefore beyond the scope of this chapter; some are discarded as they do not apply in this case; and some are instances of more general problems either with artificial insemination techniques or parenting in general.

Wilkinson starts by explaining that he is interested in questions regarding the regulation and legislation of practices relating to 'designer babies' and by setting aside considerations about the child's future welfare, as a full discussion of these is beyond the scope of this chapter. The discussion, which looks at two arguments in favour and a number against 'designer babies', is set in the background of three types of cases of reproductive choice: saviour siblings, sex selection and designer disability.

The first argument in favour of 'designer babies' appeals to a general argument in favour of respect for autonomy and freedom with respect to our choices. This general respect for autonomy can be applied in particular to respect for reproductive choices and a corollary to this argument stresses the central importance of reproduction in human lives. The second type of argument in favour appeals to practical arguments regarding the benefits of allowing people this kind of reproductive choice. These include the benefit for the children which are produced using these techniques and also the advantages for anyone else who benefits from the procedures, e.g. sick siblings. Although Wilkinson expresses some reservations about these arguments, the rest of the chapter is dedicated to arguing against two objections to 'designer babies'.

The two objections considered are: the idea that 'designer babies' involve the instrumentalisation of the child, and the argument from a right to an open future. The idea that the child may be used as an instrument for one's own purposes can be understood in two ways. It can refer to an argument that the child was selected for the wrong reasons, or to problems with how this child is likely to be treated in the future. Wilkinson focuses on the first interpretation, that the child must be wanted for its own sake.

This Kantian-inspired idea that children must be wanted for their own sake is used very frequently, but is subject to a number of objections. Fairly straightforwardly one could argue that it is a misrepresentation of the Kantian imperative, which specifically prohibits treating others *solely* as means to one's own ends. Furthermore, if we can make sense of the prohibition, then it looks like it applies to all sorts of reasons for having a child. Many of the common reasons why people have children without the assistance of technology are morally problematic in this respect, so the case against 'designer babies' is not unique. Without having to go into the complex arguments relating to whether embryos should be considered as ends-in-themselves, Wilkinson concludes that assuming this status for embryos creates problems for many of our practices at the beginning of life, practices which, unlike 'designer babies', we tend to consider unproblematic. The issue of the embryo's moral status is left to another discussion, but if embryos do have moral status, this conclusion calls for consistency in our reasoning.

If the direct argument for instrumentalisation fails, perhaps we should consider an indirect one. Perhaps allowing 'designer babies' will have harmful effects on the child, encouraging inappropriate parental attitudes. This is a form of indirect instrumentalisation. This brings us to the idea of a child having a right to an open future, such that denying a child this is wrong, in that it is a way of failing to respect the child as an end-in-itself.

We can make sense of the right to an open future by seeing it as drawing attention to the form of this right rather than its specific content. It's a right-in-trust, one that the child does not possess now, but which can be violated in advance. Wilkinson accepts that this is an appealing interpretation and one that seems to explain uncontroversial cases, i.e. cases where we want to object to parents' attempts to cut off important choices in the child's future. However, it does not seem to work as well in more problematic cases, cases where the value of the options themselves may be a matter of contention. Parents influence their children in a variety of ways; in fact, we expect good parents to have a say in the kinds of values they instil in their children. In effect, we expect parents to close off some reprehensible options, e.g. becoming a criminal. So the right to an open future should not be understood as a right to have all possible options available, but rather as having all worthwhile options available, which introduces an evaluative element in the decision. This is a point we will return to below.

Meanwhile one could object that even if we make sense of the idea of a right to an open future for children, surely this cannot apply to embryos, as they cannot have a right to a future without a right to life. One way out of this is to reject the strong version of this argument which presupposes a right to a future and the idea that if you exist, you have a right not to have certain options foreclosed, with a weak version which becomes conditional: if you have a future, you have a right not to have certain options foreclosed.

Pre-empting another objection, Wilkinson is also willing to abandon talk of rights altogether in favour of the Open Future Principle: it is wrong *prima facie* to create, or help to create, a person whose future is insufficiently open in relevant respects, or to alter a person's life in ways that cause his/her future to be insufficiently open in relevant respects. The Open Future Principle allows us to explain what is wrong about creating a life which is insufficiently open in the first place and/or causing a life to be insufficiently open. Importantly, all this can be achieved without having to decide the question of the moral status of the embryo.

We can now apply the Open Future Principle to the three cases of 'designer babies'. Sex selection in itself does not seem to limit future options. Saviour siblings may not limit options when the contribution to the sick sibling is minimal, e.g. use of umbilical cord, and if there is a problem about psychological pressure which may affect the freedom of one's choice to donate to one's family, then this is a general problem about choice; and it should be addressed as such and not as a specific objection against 'designer babies'. Finally, on the issue of selecting for disability, much will depend, as above for Lillehammer, on the particulars of the disability and how it affects one's choices. There is an empirical question about how deafness, for example, affects one's range and type of choices, but also there is a separate question about whether parents are obliged to maximise a child's choices (as in the Lillehammer superhuman case). Again, when discussing the right to an open future, one will need to evaluate the choices themselves and whether these choices are constrained in relevant and serious ways. This, however, seems like a project for another occasion. Wilkinson's chapter, much like Lillehammer's, is an invitation for further discussion, both chapters inviting us to view practical problems in a particular way, one that does not offer a quick and easy solution but is a call for further thought.

The next chapter, by Heather Draper, takes up the challenge of responding to some of the practical considerations concerning the child's future welfare which were set aside by Wilkinson. This contribution takes as its starting point a change in British law. From April 2005 there is no longer anonymity for new gamete donors. Draper takes this as her starting point to argue that while it is permissible for donor-conceived people to contact a willing donor, they have no right to do so.

Draper's argument starts by examining two ideas which have been put forward against the change in the law. The first is a worry that this change in the anonymous status of donors will deter potential donors. Whatever the empirical backing of this worry, it should be discarded, since the fundamental question is whether people are harmed by being denied access to this information and whether this harm can be redressed by doing away with donor anonymity. The second is a claim that gamete recipients are entitled to privacy. Three arguments can be made to counter this point:

1. children have a right to know the identity of the donor;
2. parents ought not to be dishonest with their children;

3. it is better for children to know the circumstances of their conception and have an honest relationship with their parents.

However, Draper points out, one can accept 2 and 3 without accepting
1. Three arguments have been put forward in support of the change in
the law and Draper considers and rejects them in turn: a) a claim of
parity between donation and adoption; b) a right to know one's genetic origins; and c) a claim about the significance of a complete genetic
history for medical treatments. In doing so she draws a distinction
between a right to honesty and a right to be able to identify one's
donors.

First, Draper considers the similarities between adoption and donation. In both cases children lack a close genetic connection to their
social parents and may try to find out more about their origin.
Infertility, like adoption, is stigmatised and veiled in secrecy. However,
the correct response to these points is to encourage a more pluralistic
account of the family. We should encourage change so that adoption
and infertility become more rather than less socially acceptable.
Paradoxically, giving children the right to know reinforces the supremacy of genetic connectedness over social connectedness and under-
values adoptive and social parents.

Furthermore, there are significant differences between adoption and
donation. Here we need to ask: if there is an interest in finding out
information about one's genetic origins, what kind of interest is it? In
the case of adoption, it could be a social need related to feelings of
abandonment, or mere curiosity, rather than a genetic need. One could
argue that there is a moral difference between adoption and donation
as adoption is the final act of parental responsibility. However, it is not
clear in what sense donors have parental responsibilities towards the
children created this way. At best one could argue that donors have a
responsibility to donate to suitable families and to respond favorably
to future requests for information. However, crucially, even this weak
argument depends on the importance of genetic relatedness (we will
return to the significance of this point below). Thus, the argument
becomes circular: the similarities between adoption and donation are
appealed to in order to explain the importance of genetic connectedness, but we can make sense of the similarities only by assuming the
importance of genetic connectedness in the first place.

Moving on, two thoughts seem to motivate the right to information
about one's genetic origins: first, the importance of genetic information for medical treatment, and second, the importance of knowing

one's genetic origins. The first is easily dealt with. Genetic connectedness *per se* cannot generate this right, so it must be a general principle
about helping those in need. Endorsing this would require us to
endorse a general duty of rescue, but this does not seem to be the
motivation behind this change in the law. Furthermore, we need to ask
what kinds of information, specifically, would be of use to the child
relevant to his need. General information which may be of use in
medical treatments is rather different from knowing the identity of
one's donor; and revealing the identity of the donor in sharing genetic
information has to be considered in the context of revealing genetic
information in general to family members.

Next, we come to the most important argument for a right to know –
the idea that it is important to know one's genetic origins. However,
Draper points out that the importance of having this knowledge is
often assumed without much argument. It is not clear why it is important to know which of our physical characteristics we share with our
genetic parents, and even appeals to a sense of belonging which is
achieved through this knowledge are unclear. Background information
about one's values, culture and religion seem to contribute more to our
sense of belonging than knowledge of our genetic parents. Crucially,
though, what this discussion seems to point to is that the entitlement
to know involves more that having a name and address. It seems to
imply that the donor person is entitled to a relationship with the
donor and access to a wider family. However, Draper's main point is
that it is difficult to see how one person can be given the right to have
this kind of relationship with another or how genetic connectedness in
these circumstances grants one such a right.

Compare donation with a case in which a child is unjustly removed
from its parents. In such a case there is a right to know. However, it is
not generated by the genetic connectedness, but rather by the unjust
circumstances under which the separation occurred. Donation does
not involve the same kind of initial injustice; donors freely donate and
the child doesn't even exist at the time of the donation to be harmed.
Now we should be careful not to confuse the idea that people have the
right voluntarily to enter into relationships, including relationships
with those they are genetically related to, with the idea that donor
people are owed a relationship with their donor because of the genetic
connection between them. Surely the obligation to provide this kind of
relationship belongs to the social parent who has brought the child up.

Having argued against the idea of a right to know, Draper makes a
final move by looking at some of the implications of accepting such a

right. One potential problem is that British law is now unjust and ought to apply retrospectively to many other similar cases. Furthermore, if there is a right to know, it should be available to all, not just donor people, making paternity tests a right for all, but surely no one wants to follow the argument to this conclusion?

Katrien Devolder and John Harris take on the demanding task of questioning the consistency of our beliefs, practices and regulations regarding the use of embryonic stem cells in research. Rather than entering the debate over the moral status of the embryo head on, they argue that even those who do think that the embryo has moral status equivalent to that of a person fail to apply this idea consistently. Given that they are willing to treat the embryo as if it does not have moral status elsewhere, and that embryonic stem cell research holds the promise of greatly benefiting actual, existing persons, one has to conclude that this kind of research is permissible and even obligatory.

Devolder and Harris start by pointing out the potential benefits of stem cell research in general, which include the possibility of treating currently incurable diseases and conditions, the possibility of restoring tissue rather than merely halting disease, and also developments which may lead to new drugs. There are three lines of stem cell research, from *in vitro* embryos, cord blood-derived stem cells, and stem cells originating from tissues or organs from fetuses or organisms after birth, referred to as adult stem cells. This chapter examines the most common objection to the use of embryonic stem cells, which is captured in the thought that the embryo is 'one of us'. This idea is understood broadly, sometimes referring to the objection that one should not use embryos as a means to an end (discussed in this form in detail by Wilkinson) or sometimes referring to the thought that research on embryos involves reprehensible killing.

Given that there is this objection specifically to the use of embryonic stem cells and there are other possible sources of stem cells, one line of thought suggests that there is no need to permit the contentious use of embryonic stem cells. Rather, the same benefits can be realised using the less contentious sources of stem cells. This idea is known as the principle of subsidiarity and assumes that there is a hierarchy of more morally acceptable methods of obtaining stem cells. However, this solution is too quick as there is still scientific uncertainty over which line of research is more promising. If it turns out that embryonic stem cells have significant scientific advantages over other sources of stem cells, then we are faced with the same difficulty.

It seems that one can embrace one of two positions on our moral obligations regarding embryonic research: either the embryo has the same status as a person, in which case this kind of research should be legally prohibited, or *in vitro* embryos, regardless of their origins, may be used for scientific research on condition that they are not then implanted in the womb. However, many laws, regulations and guidelines have embraced an intermediate position, seeking to justify some kinds of research but not others. Devolder and Harris now move on to show why such intermediate positions are not convincing.

One type of intermediate position is based on the 'use–derivation distinction'; the use of ES cells has a different moral status from their derivation from *in vitro* embryos. The former is morally acceptable whereas the latter is not. Another type of intermediate position relies on the 'discarded–created distinction'; use of spare IVF embryos for research is morally acceptable, unlike the use of embryos created solely for research. In general, counties which are currently very restrictive with regard to embryonic research may be led to some sort of compromise as the techniques will be available for their citizens elsewhere. However – and this is where Devolder and Harris make their central claim – these kinds of intermediate or compromise positions are going to be inconsistent. They go on to show that one cannot adopt such an intermediate or compromise position without being morally complicit with evil deeds.

Take the claim that using embryonic stem cells is morally permissible, but deriving them is not. For this distinction to work one has to assume that using embryonic stem cells is acceptable because they are not equivalent to embryos (and we are accepting that embryos are 'one of us' so their use is immoral). One suggestion here is that embryonic stem cells are crucially different from embryos as the former are *pluripotent* and the latter *totipotent*. However, Devolder and Harris point out that for this argument to work, one has to explain why it is ethically acceptable to use products obtained through a wrongful act (the use of embryos in the first place). Embryonic stem cells can only be obtained by killing embryos, which are 'one of us'. If we accept the use of embryonic stem cells procured from an immoral act, are we being morally complicit in this evil?

At the heart of this problem is a question regarding our responsibility for benefit received from another's wrongdoing. One way out of this is to argue for a separation principle, i.e. the act through which the cell products are obtained should be completely separated from the use that is made of these products. One could argue that in this case the separation holds as the use of surplus embryos and taking precautions to ensure that one does not encourage the immoral treatment of embryos for the procurement of stem cells in the future maintains this separation between use and derivation. However, this is implausible. The separation argument does not hold for a number of reasons. First, the line drawn is arbitrary. Furthermore, it is likely to be extended, as the source is insufficient and the benefit great. Finally, payment for stem cells does directly encourage their procurement in the future.

If no separation is possible, then support for the research is acceptable only if the process of obtaining stem cells for the research is acceptable. The claim is that the embryo is 'one of us' and that therefore the process of obtaining stem cells is not acceptable. However, our practices involving embryos contradict this claim and seem to show that we have little respect for the early embryo. Concerns for embryos tend to focus on embryos we have an interest in, i.e. embryos involved in reproductive efforts. Furthermore, we are rather complacent about the creating of embryos for in vitro fertilisation (IVF), but if it is acceptable to create and sacrifice embryos in order to create new life, why is it not acceptable to create and sacrifice embryos to save existing life?

If creating stem cells is evil and benefiting from stem cells is evil, we should reject these benefits, but we tend to accept such benefits, while at the same time allowing for evil, inconsistent treatment of embryos in other practices. Those who believe that the embryo is 'one of us' are committed to a position even they do not uphold in their practices and it is this inconsistency that Devolder and Harris object to.

The next chapter, by David Oderberg, offers a new approach to the issues raised by genetic engineering in general. To be more precise, not a new approach as such, but a re-evaluation and redevelopment of a traditional approach, one that, however, has often been summarily rejected at the start of debates in this area. Natural law theory has often been dismissed because of its, allegedly, implausible appeal to the natural/unnatural distinction. Historically, the focus of discussions in this area has been on consequentialist arguments, waged in terms of balancing the potential harms and benefits of proposed techniques and procedures. Oderberg takes up the challenge of showing that natural law theory can provide a reasoned and reasonable response to issues in biotechnology and that its opponents are not justified in summarily dismissing it.

Oderberg starts with an account of how we should *not* understand the natural/unnatural distinction. This is because often misunderstandings of the distinction lie at the heart of the objections that lead to the rejection of natural law theory. One way of interpreting the distinction is that it is open to the objection that it is trying to derive the 'normative' from the 'descriptive'. Surely appeals to nature are appeals to how things naturally are, and should not be used to derive conclusions about how things should be. However, although natural law theory is often objected to for relying on this claim, it is not evident that natural law theorists in fact do support this claim. What they do want to say about nature is that nature is a complex network of law and processes that enable adaptation for survival, and that considering it as such may lead us to appreciate what makes our lives go well.

Another interpretation of the distinction is that it appeals to a notion of the 'normal', which in turn should be understood as a statistical or probabilistic term. Critics are right that statistics don't provide independent grounds for appealing to the natural, they simply give us information about the prevalence of characteristics and practice measured. However, we can appeal to nature for evidence pointing to the fundamental characteristics of human life, characteristics which ground the ethical judgement.

Yet another interpretation sees the appeal to the natural as an appeal to God's will. The objection here is that God's will is difficult to determine, but Oderberg has a number of responses to this. Understood as an epistemological problem it is a problem for ethics in general. Understood as a free will problem, there is a long tradition of trying to respond to it which is not reflected in the hurried rejection of natural law theory in bioethics. A natural law theorist could respond that the natural order of things structures the world antecedently of human desires and preferences. Human beings discern this natural order without necessarily appealing to God, and can have access to moral truth through reflection upon the natural order.

Finally, the distinction is objected to as implausible as we are part of nature, so everything we do is natural. Surely, this interpretation is easily dismissed, as although we are part of nature we are also rational, moral beings and it is this perspective we should focus on.

Having discounted the ways in which the natural/unnatural distinction has been misinterpreted, Oderberg moves on to construct a natural law account of genetic engineering. The naturalistic fallacy charge, the idea that one cannot derive an 'ought' from an 'is', is based on a dichotomy the natural law theorist would not accept. The natural and the normative are not clearly dichotomous as required for the charge to have any force; rather the normative is somehow part of, continuous or derivable from nature. Here Oderberg uses this point to raise his own objections against the dominant consequentialist perspective in bioethics. Consequentialism encourages us to evaluate techniques and practices in terms of the good or bad consequences they bring about, disregarding their inherent nature. Not only that, but the notion of 'harm' employed here is essentially normative, as we can't make sense of it without understanding the natural order of things according to which this is a harm.

A natural law theorist would want to agree with a consequentialist over the centrality of harm, but when it comes to applying the concept of harm, the consequentialist wears 'moral blinkers'. Consequentialists have a distorted conception of just what sort of harm a practice might cause, whereas natural law theorists understand harm by looking at the conditions under which human nature is fulfilled. They ask what causes human nature to flourish or conversely to be perverted, damaged or impaired. The normative is built into the natural. Therefore, the natural law theorist will aim to elaborate on a list of basic human goods, goods which different people are free to place different emphases on and will object to practices which undermine human goods in their very pursuit.

Oderberg goes on to give us a flavour of what a natural law theorist would have to say about cloning. Practices which undermine family life undermine the good of human flourishing, so we can object to practices which cut off children from their genetic parents and therefore their sense of who they are. Cloning, unlike adoption, is objectionable because it involves secrecy and deception, but also because such knowledge is an intrinsic good, whether or not deception harms that person. Furthermore, the practice is exploitative, leads to commodification, serious mistakes, the unacceptable introduction of a market, etc. Fundamentally, the practice undermines the goods that make possible the proper functioning of society and the individuals within it. In general, a natural law theorist, when evaluating practices, wants to know what goods, if any, are undermined by it, whether the goods involved are personal or social, how they are undermined, and what the intent is of the practitioners concerned. Thus, it is not artificiality as such or technology which is rejected.

The natural law theorist has more to say on our relationship to nature. Use of nature must conform to nature's own requirements, rather than being exploitative. And these requirements are the basic goods through pursuit of which humans flourish. Evaluation of most practices will require considering them in their context, evaluating their outcomes, seeing their impact on human flourishing, etc.

Oderberg gives us an idea of what such an analysis would be like by considering cloning, 'designer babies', sex selection and stem cell research.

As might be expected, much more can be said about applying natural law theory to specific questions; however, Oderberg has provided us with the first step in any such discussions by rescuing natural law theory from some obvious misinterpretations and analysing the theory in a plausible and promising direction. This new direction requires a radical rethinking of how we evaluate practices in bioethics, but the challenge to do so has been laid down.

The concept of autonomy has, arguably, had the greatest impact in medical ethics. So much so that respect for autonomy seems to be accepted as a self-evident justification for any argument. The next chapter looks deeper into this assumption. James Taylor examines exactly what it means to be autonomous and how this conception of autonomy affects the arguments for a market in organ transplants.

Taylor starts by acknowledging that there is widespread agreement against the idea of a market in transplant organs based on a claim that this would compromise autonomy. However, he argues that a market in transplant organs would not compromise autonomy through coercion, nor would it compromise autonomy through irresistible financial inducements. In fact, a first look at autonomy may suggest that a market in transplant organs would enhance autonomy as it involves an additional option. This argument rests on two assumptions: that the vendor is autonomous in the first place, and that the option to sell is not an autonomy-compromising option. Taylor's task is specifically to defend the first of these two assumptions.

Taylor's first step is to clarify the concept of autonomy by pointing out that it is not coextensive with identification. Using the example of a person who is deceived into acting by another, Taylor makes the point that it is possible for an agent to lack autonomy (which is not compatible with deception) but still be moved into action by a desire which he volitionally endorses (one he identifies with). The crucial conclusion of this detailed and complex discussion is that for a person to identify with her first-order desires, to endorse them, is not sufficient for her to be autonomous. Autonomy requires control over one's desires and actions.

Given this clarification about the understanding of autonomy, the chapter moves on to consider two objections to a market in transplant organs: the argument from coercion and the argument from irresistible offers. The argument from economic coercion claims that economic disparities and poverty will coerce people into selling their organs. This coercion amounts to a diminution of autonomy and is therefore objectionable. However, why does coercion amount to a diminution of autonomy? It can do so only if control over the decision is ceded and control can be ceded only to another agent. Since poverty is doing the coercing it is conceptually impossible for a person to suffer from a diminution in autonomy as a result.

The argument from irresistible offers can be understood to be making a variety of claims, some of which are easier to reject than others. One interpretation is that an irresistible price will be offered for the organ, but this is unlikely. Another interpretation is that creating a market in organ transplants will lead to weakness of will, with agents unable to resist offers. However, this point is true of all markets for any good and doesn't apply specifically to transplant organs. A third interpretation may refer to the idea that to sell is irresistible in that other options are rendered ineligible. Again, however, this is based on a misunderstanding. This argument supposes that the more options one has, the more one is able to exercise one's autonomy, but this conflates the deliberative aspect of autonomy with its exercise. Finding an option so appealing that one can endorse it immediately is not evidence of a diminution of autonomy.

A fourth way of understanding the argument from irresistible offers suggests that ambivalence precludes autonomy. Genuine ambivalence between two options means that there are no good reasons for favouring one over another, so choosing either in such a case cannot be a diminution of autonomy as without an identifiable ground for choosing there can be no exercise of autonomy. For this argument to have force one has to see autonomy as intrinsically valuable, which brings Taylor to consider what the value of autonomy is.

It seems that people value autonomy in itself, even when giving up the right to choose to someone else may have resulted in a better outcome in terms of the choice. However, crucially, we value autonomy not as a means to our wellbeing but as a part of it. If the value of autonomy is derivative from that of wellbeing, then it is not intrinsic and the fourth interpretation from irresistible offers cannot be grounded. If we are to decide on markets for transplant organs, we should do so based on considerations of wellbeing rather than autonomy.

Empirical studies into the effects of organ sales seem to indicate a significant drop in wellbeing in vendors, which would suggest that prohibitions are well placed. However, we have to be careful how we interpret these results. A more careful look might reveal that the adverse effects on wellbeing may be the result of the prohibition itself, which forces people into a black market, with all the repercussions in terms of secrecy and no recourse to the law such a market has. Therefore, concern for the vendors' wellbeing should lead us to legitimise, regulate and control markets for organ transplants.

If discussions of autonomy have been central to medical ethics research, they have been almost exclusively focused on the autonomy of patients. The next chapter, by Piers Benn, changes this focus as it discusses the problem of conscientious objection. Broadly speaking, the question of how much leeway should health care professionals be allowed in following the dictates of their conscience when these conflict with their professional obligations can be seen as an extension of the issue of professional autonomy.

Benn starts by reflecting on some of the ways we speak about following one's conscience and in particular how we evaluate this as something particularly noble, especially when done at a cost to the agent. This raises the question of whether there is anything admirable about following one's conscience over and above what is admirable about doing the right thing.

A conceptual analysis of the term 'conscience' reveals three features. The first is that following one's conscience is necessary but not sufficient for doing the right thing. Following one's conscience is not in itself a sufficient excuse for doing the wrong thing. One must do what is right, think that it is right and do it because one thinks it is right. Second, appeals to conscience are self-addressed. This should not be interpreted as a relativistic claim, but rather that it is wrong for anyone to do what he believes is wrong. In addition to doing what our conscience dictates, we also have an obligation to ensure we acquire a sound conscience. Finally, there is a claim that conscience be understood as a source of moral knowledge, one perhaps even more worthy than authority. This claim requires interpretation. If seen as a claim about consulting one's conscience for determining what is the right thing to do, then it really seems to be about deliberating. In which case we should we wary of the fallibility of our judgements and of awarding our judgements authority simply because they are our own.

Having said something about how we should understand conscience, Benn goes on to ask what is wrong with doing something one holds is wrong. Consider a case where an agent does something in the mistaken belief it is wrong – has he done anything wrong? According to a plausible account he has, as he has shown insufficient concern for morality; agents should not be excused when doing what they think is wrong even when that happens to be right. To understand this better we need to look at the role of moral conviction.

According to an internalist account of motivation, holding a moral conviction leads to acting in accordance with it. Although there are some cases where people act against their conscience, namely weakness of will, it is difficult to make sense of someone who systematically disregards the dictates of his conscience. If conscience cannot be systematically motivationally inert, how is it possible for people to do what they think is wrong? One answer is hypocrisy, which may involve a pretence at morality or self-deception that one is acting in accordance to one's conscience.

Another answer is that the agent is committed to certain values but not certain what her real principles are. In cases where people act against their conscience we might want to say that they changed their mind, they acted in conflict, they acted wrongly, but we might also want to say that while they retain the idea that an act is forbidden they still feel a pull towards certain of its features. The example of a nurse who is committed to opposing euthanasia, but is sympathetically drawn towards helping alleviate a patient's pain by helping him die, makes this point. Similarly, we should interpret the example of Huck Finn as someone who is motivated by sympathy as a virtue even when this goes against his official conscience. Huck's role in the escape of a slave should be seen as evidence of his real conscience and since he is unable to ignore his feelings of sympathy in spite of what he claims to believe morally to the contrary we should see this as a sign that his deep dispositions are appropriately sympathetic. Further reflection may lead the nurse and Huck to reconsider their conscience.

To return to the original question: What is wrong with doing what one thinks is wrong? Professional roles include allegiance to certain duties integral to the profession; however, having undertaken these duties agents may find that they conflict with the dictates of their conscience. At the same time, it seems that someone is harmed or wronged when forced to do what he thinks is wrong. But in what way is the agent harmed or wronged? One suggestion is that it is distressing to do what one thinks is wrong. This is a good reason for not forcing people to do what is wrong, but not good enough to defeat their professional obligation to do it anyway. Perhaps a better account can be given if we look at the kind of distress that is caused. Going against the dictates of one's conscience involves not only doing something wrong, but *me* doing something wrong. In a sense being forced to do what one thinks is wrong is an attack on one's integrity. Our values define who we are and forcing someone to act contrary to his values is forcing someone to act contrary to himself. In normal cases, excepting extreme evil, we can appreciate the values those we disagree with are trying to uphold. Respect for the conscience of those we disagree with is more often founded in the knowledge that the people we disagree with do care about morality and – crucially – have some sense of basic, genuine, moral values. If we can make sense of this idea, we can make sense of what is wrong with forcing someone to act against their conscience.

One final observation is relevant here and this is that certain professions have widely known and shared core values and practices. For this reason, entry into the profession requires some kind of fundamental agreement with these values. Of course, there may well be disputes about how these values are instantiated or how to resolve conflicts between them, but those who conscientiously object to them in a fundamental way can reasonably be refused entry into the profession.

The last two chapters consider issues raised by the possibility of euthanasia. Nafsika Athanassoulis considers the distinction between killing and letting die; whereas Ray Frey uses arguments based on analogies to show that the causal role of doctors in withdrawal of treatment is similar to that in physician-assisted suicide and active voluntary euthanasia.

The starting point for Athanassoulis is a point of law. British law as it stands has evolved through case studies to suggest that decisions on whether to withhold/withdraw treatment for incompetent minors should be based on a judgement about the quality of their lives. Given that we have to make this kind of decision, it is possible in some cases that the patient's quality of life is so bad they are better off being allowed to die. Such decisions are restricted to cases of allowing patients to die.

Using the example of Charlie, a neonate with disabilities and in need of life-saving treatment, and Britney, a normal baby in need of the same treatment, Athanassoulis makes the point that what motivates the decision not to treat is a judgement about the patient's poor quality of life coupled with an understanding that death is not the worst thing that can happen. At this point supporters of the doctrine of the sanctity of life may object to the moral permissibility of such a course of action, and this could be a significant objection. Little more can be said on this other than this kind of interpretation of the doctrine of the sanctity of life requires one to see life as an absolute and incommensurable good, and this conception must be applied consistently in all sorts of different cases. A further objection that Charlie's case should be characterised as an example of futile treatment is rejected on the grounds that the treatment itself is very successful, and if anything, a judgement is made that Charlie's *life* is futile.

Suppose we now consider Douglas, a neonate similar to Charlie in terms of quality of life, but who does not contract an easily treatable, but potentially fatal condition. Douglas's options are different from Charlie's as the choice whether to treat him or not never arises. The main argument of this chapter is that the difference between Charlie and Douglas is entirely down to luck and therefore should not be allowed to affect the kinds of options available to these patients. A similar argument can be made on behalf of competent patients who cannot, for reasons outside their control, commit suicide, or refuse treatment and require assistance to be killed.

Central to this argument is the idea that sometimes it is inappropriate to allow the kinds of choices that are available to us to be a matter of luck. Important decisions about who dies and the manner and timing of their death, like decisions regarding health care allocation, should not be beyond our control, i.e. left to luck. Controlling the influence of luck is also a demand of justice in this respect and the charge of discrimination can be made against practices which restrict options for people simply because they are disabled.

Of course, a chapter on this kind of topic has to say something about distinctions which have played a central role in this debate; namely, the difference between acts and omissions and the importance of intentions. This part of the discussion starts with a clarification about euthanasia. Euthanasia differs from accidental deaths as it is intentional, but differs from murder as the motive for it is beneficent. It is important not to prejudge the question of the difference between killing and letting die by assuming that the motives in the former practice are always malevolent while those of the latter benevolent. But what of the distinction between killing and letting die itself?

One of the reasons why letting die is presented as a fairly innocuous practice is the claim that the agent has not actually done anything. However, here it is important to distinguish between non-actions and omissions. Omissions have to be understood within a context which requires action, and therefore whether they are justified or not depends on whether this requirement for action is defensible. Furthermore, omissions interpreted like this are part of agency and we can be held responsible for them, especially when seen within a context within which action is expected of us, e.g. the role of doctors. Lastly, omissions can be crucial in bringing about the outcome and this is different from the idea that omissions do not affect the status quo. What all this means is that we cannot judge an omission to be acceptable simply because it is an omission, so letting die cannot be morally permissible in comparison to killing simply because it involves an omission.

Given that the doctor's intentions in letting Charlie die and in killing Douglas are benevolent, this is what makes both practices morally acceptable and restricting Douglas's options simply because of an element which is down to luck, i.e. he happened not to require life-sustaining treatment which could be withheld, is unfair.

Finally, one last objection is considered, which is the claim that in Charlie's case there is no intention to bring about the death and this is a morally significant difference between his case and killing. Here we might want to ask whether the doctor would proceed with the omission of Charlie's treatment if he knew the death would not occur and the answer to this is no, as not treating would simply worsen Charlie's quality of life and not be in his best interests. The decision not to treat makes sense only if seen as involving the intention to bring about the death. One can intend the death and see it as a benefit for the patient even if whether this death comes about or not is not within one's power. We should, then, eliminate the influence of luck and if this involves owning up to the kinds of decisions we make, decisions about quality of life and the nature of death and at the same time we acknowledge that there is no significant moral difference between killing and letting die in such cases, then killing patients, in these circumstances, is as acceptable as letting them die.

Ray Frey's chapter begins by drawing some distinctions and clarifying some definitions. Similar to Athanassoulis, euthanasia is defined as intentionally bringing about death from benevolent motives and generally applies to patients who are gravely and terminally ill, although their lives need not be in danger at the point of considering euthanasia. Frey interprets voluntary requests for euthanasia as being based on judgements about one's quality of life.

He then considers a number of distinctions which are often raised in this debate. First, many instances of passive euthanasia involve an action, such as switching off a ventilator, so it is difficult to see how they differ from active euthanasia; furthermore, given that the doctor has knowingly carried out actions it is difficult to claim that the underlying disease actually killed the person. Second, in practice, cases of non-voluntary euthanasia may become cases of involuntary euthanasia. Third, a difference between physician-assisted suicide and active euthanasia is that in the former the last causal actor in bringing about the death is the patient, whereas in the latter it is the doctor.

Many of the objections against the legalisation of euthanasia are concerned with some variation of the slippery slope arguments. Concerns are raised about slopes leading to involuntary euthanasia, leading to incompetent patients unaware of what they are requesting, leading to all sorts of methods of killing patients and leading to patients being pressurised to agree to die. However, in the absence of concrete evidence that such slopes will occur and the absence of proper discussions on the feasibility of safeguards, Frey remains unimpressed by slippery slope worries. Importantly, we must remember that for the critics of euthanasia it is not the end consequences of such slippery slopes which give rise to such opposition, but rather the impermissibility of the original practices themselves.

Having set the scene within the debate on euthanasia, Frey concentrates on the idea that patients have a right to refuse treatment. Such a right seems important since without it we would be forced to live lives we did not want, lives which had been judged to be worthwhile by others. This is not just an attack on the individual's autonomy, but involves a doctor taking moral control over one's life. Frey argues that we should understand the withdrawal of treatment as a kind of suicide, since it is done in the knowledge of the consequences of the refusal and involves the intention to die. However, if refusing treatment is morally acceptable and is an instance of suicide, why is it acceptable to commit suicide but not acceptable to allow physician-assisted suicide? If choosing to die is morally permissible, why isn't it permissible to supply others with the means to die?

A claimed difference between such cases is the worry that the doctor becomes the cause of death. However, the doctor is the cause of death in acceptable cases such as the removal of a ventilator or in over-medication which results in death, without this fact about the causal links leading to death actually conferring guilt on the doctor for that death. All we can conclude from this point is that what one causes in the world is relevant to what one is morally responsible for. Both the doctor's intentions and the fact that the patient requested to die go towards the moral permissibility of the act, but neither of the two shows that the doctor did not cause the death. The important conclusion of this discussion, then, is that the doctor is more closely connected with the patient's death in cases of withdrawal than in cases of physician-assisted suicide, but that not much turns on this fact alone. There is no difference in the causal structure of withdrawal on the one hand and physician-assisted suicide and active voluntary euthanasia on the other; therefore, this cannot be the main reason why the former is morally permissible and the latter is not.

1
Benefit, Disability and the Non-Identity Problem

Hallvard Lillehammer

Preconception and identity

It is natural to think the evaluation of reproductive decisions is subject to the same ethical standards that apply to relations between existing persons. If so, prospective parents should be able to extrapolate from the latter to the former when thinking about whether to have children, how to have children, what sort of children to have, and so on. Yet there are well-known features of certain reproductive decisions that make it hard to grasp how some of the most basic ethical standards that apply to relations between persons also apply to them. These features obtain in scenarios where reproductive decisions are made in the absence of any distinct or identifiable person who fills the role of primary beneficiary or victim. I call such scenarios *pre-conception scenarios*, and any scenario where the causing to exist of an entity is at stake a *non-identity scenario*. The problem of how to evaluate decisions ethically where the identity of the entity affected is itself determined by those decisions is sometimes called the *non-identity problem*.[1] I shall follow this usage. Pre-conception scenarios form a subset of non-identity scenarios. This chapter is primarily about the non-identity problem as applied to pre-conception scenarios, although I also discuss a number of other non-identity scenarios. I shall not attempt to solve the non-identity problem here, either as it applies to pre-conception scenarios or elsewhere. What I do hope to achieve, however, is to shed some light on pre-conception scenarios by locating them with respect to other non-identity scenarios where the distinctive features of human reproduction are absent. For agents also face non-identity scenarios where the entities created are not even potential holders of the interests or rights that ethically constrain our behaviour towards future or potential persons. By locating pre-conception scenarios in this wider context, I will suggest that traditional discussions of the non-identity problem have taken an overly narrow view of the ethical concerns that govern decisions in pre-conception scenarios. In doing so, I shall make use of a contrast between two distinctions, namely the distinction between person-involving and non-person-involving considerations on the one hand, and between partial and impartial considerations on the other.[2] In the penultimate section, I apply this contrast to the case of human disabilities. I shall argue that the traditional approach to the non-identity problem generates a questionable view of the ethics of causing disabled people to exist.

Pre-conception scenarios have been subject to much recent philosophical controversy.[3] One explanation for this is the intimate connection between thoughts about benefit and thoughts about effects on identifiable individuals. It is natural to think that if benefit is produced, there is some distinct and identifiable individual who is benefited. This thought is sometimes called the *person-involving principle*.[4] The person-involving principle entails that where no distinct individual is identifiable as the recipient of benefit, no benefit can be produced.[5] The person-involving principle may be thought to suggest that only the interests of already existing persons are ethically relevant in pre-conception scenarios. Such scenarios include cases of in vitro fertilisation (IVF), where fertilisation of eggs by sperm takes place in a laboratory and embryos are implanted in the womb for gestation. They also include less controversial cases of reproductive choice, such as choices of reproductive partner or the time and frequency of conception, where in each case the decision made will affect the identity of any future persons caused to exist. Pre-conception scenarios are therefore sometimes proposed as an ethically more secure context than pre-natal scenarios in which to screen out harmful or non-beneficial characteristics of future individuals, precisely on the grounds that in such scenarios no person is either directly harmed or benefited by the decision to not conceive.[6]

While prior to conception there is no distinct and identifiable individual to benefit from conception, it remains true that were a child to be conceived, it would have been made to exist and its existence would take a certain form. Thus, if conception takes place between biological parents all of whose ancestors have blue eyes, any resulting child is likely to have blue eyes. Even if prospective parents were unable to discuss coherently whether it would be a benefit to any child of theirs to be conceived, they could still coherently discuss whether it would be a benefit to any child of theirs to have one type of life rather than another. Thus, any child would arguably benefit from living a highly pleasurable life rather than an intensely painful one. Such evaluations form an uncontroversial part of ethical thought and give content to the idea that it is possible for a person to have a life that is better or worse in some respect or other. Such evaluations are non-controversially available in reproductive scenarios to guide prospective parents in their choice of future offspring.

In light of this, one proposed solution to the non-identity problem is to use impartial, non-person-involving thoughts about better and worse lives to decide whether or not to conceive. Thus, in his discussion of the non-identity problem, Parfit considers the thought that in cases where no individuals are antecedently identifiable as the holders of rights and interests affected by our action, our thinking should be guided by what I call *the beneficence principle*.[7] According to the beneficence principle, where we can choose between producing more benefit rather than less, we should produce more. Thus, in pre-conception cases, prospective parents should act so as to produce as much benefit for any prospective children as possible.[8]

Parfit rejects this proposal because he thinks it entails what he calls *the repugnant conclusion*: 'For any possible population ... all with a very high quality of life, there must be some much larger imaginable population, whose existence, if other things are equal, would be better, even though its members have lives that are barely worth living.'[9] The repugnant conclusion is repugnant because it ignores the separateness of persons and the intimate relation between benefits and the individual lives in which these benefits are realised. Thus, if producing benefit is all that matters, it does not matter how that benefit is distributed within or between lives.

In response to the repugnant conclusion, some writers[10] restrict the beneficence principle to what I call *same number scenarios* in accordance with what Parfit calls the principle Q: 'If in either of two outcomes the same number of people would ever live, it would be bad if those who live are worse off, or have a lower quality of life, than those who would have lived.'[11] According to Q, the distribution of benefit is restricted to equivalence classes of individuals. It thereby rules out cases where one scenario is judged better than another by including a higher total sum of benefit shared out among a much larger group of individuals. At the same time, Q allows prospective parents to think they should maximise benefit when they bring individuals onto existence. Parfit argues that Q may be extensionally adequate. Yet he stops short of endorsing it because he thinks an account of the non-identity problem should entail Q as a special case of a more general principle.[12] Having also rejected the unrestricted version of the beneficence principle, Parfit concedes temporary defeat in his search for a general principle to solve the non-identity problem (his 'Theory X').

Even if extensionally adequate, same number principles like Q are of limited use. Standard non-identity scenarios include that of a couple who for reasons of the burdens involved would only have one child if their first child suffered from a severe handicap, but who would have several children if their first child were healthy. They include the case of a couple virtually certain to have a disabled child however they conceive and whose only alternative is not to have any children. They also include the case of any couple using IVF and risking multiple births because more than one embryo is implanted in the womb.[13] Given the plethora of cases where we cannot assume that numbers are the same, the practical scope of principles like Q is limited. It is also unclear what illumination is gained by simply replacing the unrestricted beneficence principle with a restricted, and arguably *ad hoc*, principle like Q.

Like the beneficence principle, Q is an impartial, maximising and non-person-involving principle. Thus, in any same number scenario, Q is blind to the distribution of benefit across individuals. The only legitimate ground of ethical criticism is failure to maximise benefit within the relevant equivalence class. Even if it avoids the repugnant conclusion, Q is therefore problematic for the same basic reasons that the unrestricted beneficence principle is problematic. On these grounds, McMahan rejects Parfit's optimistic statement that although 'I failed to discover X, I believe that, if they tried, others could succeed'.[14] According to McMahan, Parfit's mistake is to assume the solution must take a purely non-person-involving form. Instead, McMahan proposes a mixed view that he calls *the encompassing account*. On the encompassing account, both person-affecting and non-person-affecting considerations apply in reproductive scenarios, but in a non-additive way. In cases where person-involving principles apply, only they are ethically relevant. Thus, in pre-natal scenarios where there is a distinct individual to play the role of primary recipient of benefit or harm, non-person-involving principles like Q or the beneficence principle are inapplicable. In these cases, ethical decisions must be guided by considerations of individual rights and benefits. Thus, McMahan can say that it is wrong to cause disability in an existing foetus. In cases where only non-person-involving principles apply, only they are ethically relevant. Thus, in pre-conception scenarios where there is no distinct and identifiable individual to play the role of recipient of benefit or harm, ethical thought should be guided by impartial, non-person-involving principles like Q or the beneficence principle. Thus, McMahan can say that it is wrong to select for disease or disability using IVF.

It is unclear whether McMahan's account is a significant improvement on Parfit's approach. First, we should be suspicious of the claim that non-person-involving considerations are ethically irrelevant in scenarios where person-involving principles apply. McMahan gives no argument to rule out the conflicting view that non-person-involving considerations are ethically relevant everywhere, even if in some cases they are not overriding. This conflicting view is consistent with McMahan's reasons for moving beyond Parfit's purely non-person-involving approach. Second, the encompassing account entails a purely impartial treatment of non-identity scenarios. Even if some impartial and non-person-involving principles like Q avoid the repugnant conclusion, the encompassing account remains vulnerable to problems about the distribution of benefit across individuals within an equivalence class. Any impartial principle focused on maximising benefit, whether person-involving or not, will conflict with partial ethical concerns involved in non-identity scenarios and elsewhere. Thus, many prospective parents would not be happy to constrain their reproductive decisions by maximising benefit impartially within an equivalence class. True, some parents might base their objections on an incoherent belief in the existence of a distinct and identifiable individual at preconception stage. As I shall argue in the next section, however, this is not necessarily the case. The apparently exclusive choice between person involving principles on the one hand, and impartial, non-person-involving principles like Q or the beneficence principle on the other is an illusion. Just as with person-involving considerations, some non-person-involving ethical considerations are essentially partial. Standard treatments of the non-identity problem have failed to pay sufficient attention these considerations.

Creation and value

Impartial and non-person-involving principles like Q or the beneficence principle are based on a plausible thought. It matters ethically whether our actions make things better or worse. But what exactly does this mean in any particular case? How does it apply in pre-conception scenarios? Is it the only thing that matters? In the present section, I argue that an exclusive focus on impartial ethical considerations encourages a misleading picture of non-identity scenarios. An adequate understanding of these scenarios requires a greater sensitivity to contextually specific and partial values. It is a corollary of my argument that different non-identity scenarios call for different treatment. Thus, in Parfit's original discussion, one of his main concerns was justice between generations.[15] The problem there is how ethically to evaluate the actions of the earth's present inhabitants given their effects on the identity and lives of the earth's future inhabitants. Elsewhere, Parfit is concerned with preconception scenarios.[16] The problem there is how to evaluate ethically the actions of prospective parents with respect to their future children. While closely related, these contexts of evaluation are also ethically quite different. The future generations case is an iterated problem of social policy faced by entire populations. In such cases, there is a strong argument for maintaining a highly impartial evaluative perspective (even if it remains unclear exactly what this means). Standard pre-conception scenarios are different in at least three respects. First, they do not necessarily iterate. Second, the alternative possible populations are too small to generate the repugnant conclusion. Third, prospective parents inevitably have a highly partial evaluative perspective on their predicament. It is therefore doubtful whether non-identity scenarios present the same kind of issues in standard preconception cases as it does in standard intergenerational justice cases. In what follows, I discuss a series of different non-identity scenarios in order to bring out the partial values that characterise the choices they present.

A. The tool case

Consider a hammer. I want to put down some floorboards in my boathouse in the spring after the ice has broken up the floor during the winter. I make a hammer out of a piece of rectangular steel and a bent piece of wood I find in the shed. The piece of wood is heavy and large, and poorly suited to lie in the hand. The action of the hammer is imprecise. With lots of effort I get the work done. Had I chosen the other piece of wood in the shed I would have made a better hammer. The other piece is leaner, straight and not so heavy. Its action would have been precise. Had I chosen the other piece of wood I would not have made the hammer I made. The hammer I made owes its existence to my choosing the piece of wood I did. Was I wrong to make the hammer I made?

If asked for advice, any sensible person would say I should have used the other piece of wood and made a better hammer. It would have been better had my hammer never existed, even if it got the job done. Unlike a person, the hammer has no rights or interests, either in existing or being treated a certain way. My relation to the hammer is purely instrumental. While a hammer could in principle become a locus of intrinsic value for sentimental or other reasons, my hammer is just a bad hammer. It can be justifiably discarded if it is no longer needed and no harm is caused in the process.

To the extent that hammers exist for our benefit, their creation may seem like a perfect candidate for purely impartial evaluation. If so, that should raise our suspicion regarding the application of such principles to individuals with rights and interests like persons. Yet even with a hammer it is not obvious that I am subject to justified criticism for not maximising benefit impartially. It is my hammer. It is made from materials I found in my shed. Its existence is understandable primarily in terms of the essentially partial project of an individual fixing their boathouse. It does not follow that impartial considerations are completely irrelevant. A good hammer no longer needed is a good candidate for sharing with others or giving to charity. Even so, my reasons for making a different hammer in this case are not plausibly exhausted by the fact that by doing so I would produce more benefit impartially and non-personally considered.

B. The gift case

Consider Salvador, an accomplished surrealist painter. For the birthday of his friend and admirer Pablo, Salvador paints a realistic depiction of Pablo's favourite scene, the Grand Canal of Venice. The picture is a competently painted imitation in the style of Canaletto. Pablo is delighted. Yet the painting is not a prime example of Salvador's art. It will never have as much worth as his surrealist work. The original Canaletto is a better representation of the Grand Canal. Had Salvador produced a surrealist piece, the Canaletto imitation would never have existed. Was Salvador wrong to produce the Canaletto imitation for Pablo's birthday?

In this case, it is not clear that every sensible person would say that Salvador should have made a different painting. While neither the gift case nor the tool case involves the creation of an individual with rights or interests, the evaluation of their creation is distinct. Thus, while gifts, like works of art, may in some sense exist purely for our benefit, our relation to them transcends the purely instrumental relation in which we stand to mechanical tools such as hammers. Gifts carry an expressive value that characterises a partial relationship of respect among specific individuals. Gifts are valued intrinsically and not just for the instrumental benefits they offer. Thus, gifts, relics and other personal or culturally significant items are often among the objects that persons hold onto with the greatest effort in times of crisis and trauma.

The creation of gifts is an implausible candidate for purely impartial evaluation. The Canaletto imitation might be suboptimal in virtue of not being as good a work of art as possible. It may also be suboptimal in virtue of not producing as much benefit as possible impartially considered. It does not follow that either Pablo or anyone else is justified in complaining that a surrealist original would have been a better gift than the Canaletto imitation. Of course, Salvador could be criticised if he thought that the imitation would provide his friend with a better investment. He could be criticised if he made the imitation to frustrate Pablo's expectations or to deliberately waste his own talent. Yet none of these possibilities is an essential feature of Salvador's making the Canaletto imitation. It is therefore not obvious that Salvador was wrong to make the painting he did.

C. The pet case

Consider Eric, a black and white puppy. Eric was bred from a father and mother of pure lineage. Like many pets, Eric is not well equipped to survive in an unprotected environment. He is prone to pick up diseases, a bad hunter and afraid of things large, noisy and threatening. In spite of his idiosyncrasies, Eric is an object of great affection. He is also happy. Had the breeders made Eric's mother mate with a different dog, they could have bred a more resistant pet. In that case, Eric would never have existed. Were the breeders wrong to create Eric?

As with the gift case, it is not clear that every sensible person would respond negatively to the pet case. Unlike a hammer or a painting, a dog is a sentient being. Even if sentient beings are not holders of rights, they are plausible objects of benefit and harm. To this extent, the evaluation of their creation transcends the limits of our thinking about intrinsically valuable non-sentient objects like paintings. Pets are not merely valued intrinsically for the experience they offer. The non-instrumental relations we stand to them are partly determined by what is in their interest.[17]

The creation of pets is another implausible candidate for purely impartial evaluation. From an impartial perspective, the creation of a dog like Eric who is disabled relative to a wide range of environments is arguably suboptimal. Breeding a stronger and more resistant dog might have produced more benefit impartially considered. Had the breeders created Eric for a life in the wild, they would have exposed him to probable harm. But Eric is a perfectly normal pet. If it were wrong to create Eric, it was probably wrong to create most of the pets that actually exist. Yet for many humanly bred animals there is no such thing as their natural, as opposed to their original, environment. What matters to the evaluation of Eric's creation is not whether he is disabled relative to some arbitrary range of natural environments, but whether his existence is compatible with a good life in the environment into which he will actually be born and in which he is likely to live. It also matters what reasons the people who breed him and rear him have for wanting a dog of this kind. These reasons are likely to be as partial as the reasons people have for choosing a particular gift for a friend. It is not obvious that such reasons can be captured in purely impartial and non-person involving terms. It is therefore not obvious that his breeders were wrong to create Eric.

D. The marriage case

Javi and Pilar are each looking for a spouse. They meet on a singles cruise and fall in love. After twelve months of courtship they marry. Within three years they have bought their own house and are thinking of having children. While Javi and Pilar remain completely devoted to each other, the differences between their social backgrounds are a constant source of tension. Javier works with his four brothers in the local dockyards. Pilar is the third generation among her family to teach at university. Had Javi and Pilar not met they could each have found another partner whose background would have been a lesser source of tension. If so, their marriage would never have existed. Were Javi and Pilar wrong to get married?

Different people could reasonably disagree about the wisdom of Javi and Pilar's marital union. The issue concerns the value of a plural entity (a family or marriage) and the benefits it confers on its constituents. Plural entities (as opposed to the individuals that compose them) may not be genuine holders of moral (as opposed to legal) rights or interests. Yet plural entities are intelligible objects of intrinsic value and respect. The personal relationships that exist within a family transcend the pragmatic relations of mutual advantage and replaceability that characterise relations between strangers. Failure to respect the integrity of a marriage is a possible source of hurt and complaint. A partnership of marriage, as opposed to a partnership of convenience, is not reducible to a pure instrument of mutual benefit.

The evaluation of a marriage is also a poor candidate for purely impartial evaluation. From an impartial perspective, it may look as if Javi and Pilar should never have got married. The world is big enough for both of them to have met a different partner whose looks, intelligence, values or prospects were more in tune with their own. Yet Javi and Pilar fell in love. Love does not consist in maximising impartial benefit. Like friendship, love entails that some impartial considerations are screened off from practical deliberation. It is logically possible that the world would be a better place impartially considered if people no longer fell in love. It does not follow that the partial values on which love is based should be rejected. Thus, it is hard to believe that Javi and Pilar were simply wrong to get married, even if their decision to do so was impartially suboptimal. Indeed, some people might consider their union especially admirable in light of the social obstacles they have decided to face together.

E. The superhuman case

Jack and Jill are trying for a child. Their GP offers Jack a drug which, taken before and during the time of conception, will alter the genetic make-up of his sperm so as to make any child conceived from that sperm enjoy superhuman intelligence. Jack and Jill turn down the offer, conceive normally, and nine months later give birth to a normal, healthy child. Had Jack and Jill accepted the GP's offer they would have had a much more intelligent child. If so, their actual normal child would never have existed. Were Jack and Jill wrong to have a normal, healthy child?[18]

There might be room for considerable uncertainty and disagreement about cases like the superhuman case. Yet even if we were to approve of having a superhuman child, we would not thereby be logically committed to condemn Jack and Jill for having a normal one. In fact, most people would probably think the burden of proof is on those who favour having the superhuman child to justify their decision to deliberately 'breed' a superhuman being. Thus, according to Adams: 'the principle we all confidently endorse is not that it is wrong to bring about ... the procreation of offspring less excellent than could have been procreated, but that it is wrong to bring about ... the procreation of a human offspring which is deficient by comparison with normal human beings'.[19]

In response to Adams, it might be argued that a person of above normal human intelligence is in no way deficient by comparison with normal human beings. On these grounds, it might be considered at least permissible to have the superhuman child. On the other hand, a healthy normal human child is not deficient by comparison with a normal human being either. So by Adams' principle, it is consistent with Jack and Jill being beyond ethical criticism that they decide to have a normal child. The situation might conceivably change, of course, if everyone else were suddenly deciding to have superhuman children. Like all prospective parents, Jack and Jill are making their reproductive choices in a particular socio-historical context. This context, and the ways in which other prospective parents decide to cope with it, is bound to influence what counts as an ethically defensible decision about what sort of children to have. (Thus, it might be a relevant consideration whether a superhuman child would feel alienated from his or her 'lesser' peers, for example.)

Partly for these reasons, Jack and Jill's decision to have a normal child is in tension with impartial principles like Q and the beneficence principle. From the perspective of such principles, it might seem obvious that normal, healthy children should never be conceived if superhuman children could be caused to exist instead. Yet even if there are values that can be promoted by having superhuman children, there are equally values that the selecting away of normal human beings will undermine. Thus, the very existence of many of the aspects of their social world that Jack and Jill value most is premised on the fact that in their contingent historical circumstances most normal parents are happy to have normal children. To respect these aspects of the social world does not amount to fetishising normality for its own sake. As shown in the gift case and the pet case, it is possible to reasonably maintain an attitude of intrinsic valuation toward objects which fall short of perceived perfection in an indefinite number of ways. Likewise, it is possible to reasonably maintain an attitude of respect towards similarly imperfect objects valued by others. It is therefore not clear that Jack and Jill were wrong to have the child they did.

F. The deaf case

Consider Sharon and Candace, a deaf couple unable to conceive naturally. Both being successful health professionals, Sharon and Candace have access to expensive private IVF treatment. Using a sperm donor with five generations of deafness in his family, Sharon and Candace successfully conceive and bring to term two congenitally deaf children. If Sharon and Candace had chosen a normal sperm donor, they could have had a hearing child. But then their actual children would not have existed. While both children are congenitally deaf, they are otherwise healthy. Were Sharon and Candice wrong to have deaf children?[20]

Many people would say (and do say) that the decision of Sharon and Candace is wrong. Children are paradigm holders of interests, and the act of causing someone to be disabled might appear to constitute a paradigm example of harming someone's interests. As persons, the children of Sharon and Candace are also paradigm objects of the kind of non-instrumental valuation which gives rise to the Kantian dictum that one should never treat another only as a means, but always also as an end in itself.[21] Deaf people selecting for deafness might appear to constitute an obvious breach of this widely accepted moral principle. On the other hand, a contrary decision on the part of Sharon and Candace would entail that two worthwhile lives would never have been. While deafness is a serious disability, there is reason to believe that the children of Sharon and Candace would rather be deaf than not have existed at all. The case is therefore not as clear-cut as it may first seem.

Candace and Sharon's decision to have deaf children is in tension with purely impartial principles like Q and the beneficence principle. From an impartial perspective, it might seem obvious that disabled children should never be conceived if this could be avoided without causing harm to others (for some parents, the choice is one between a disabled child and no child at all). The denial of this claim would seem to imply that it is at least as good to be disabled as to be not disabled. Given the costs involved in meeting the special needs of the disabled, many people would find this claim hard to accept. Yet the impartial framework on which this argument is based is open to challenge. Like all prospective parents, Sharon and Candace make their choice in a particular personal and historical context. Like all prospective parents, they make that choice partly in light of partial ethical concerns. Thus, Sharon and Candace would like their children to grow up and flourish as members of their own community. If the only reason for having a deaf child were that it would make their parents feel exciting or original, that would be open to criticism.[22] Yet while Sharon and Candace are pleased that their children are deaf, their reasons are not superficial. Thus, they are concerned that their children grow up in an environment where they are able to communicate with their peers without feeling alienated. The deaf community, where the standard form of communication is sign language, arguably constitutes such an environment. An obvious precondition of this claim is the fact that the socio-economic position of the deaf community is a strong one. In fact, the children of Sharon and Candace enjoy educational and career opportunities that are above average for people in their society.[23] Sharon and Candace do not claim it would have been better for their children to be deaf in the jungle. Like all parents they are planning a family in a specific environment characterised by specific risks and uncertainties. Thus, it is ethically relevant whether their children's interests are likely to be seriously threatened by deafness. This will depend on the long-term stability of their community, the likelihood that they will find themselves unprotected outside that community, the likelihood that they will ever want to live outside that community, the extent of discrimination against the disabled in society at large, and so on. Analogous risks apply to all children, whether disabled or not. While all ethically serious parents have a responsible attitude to risk, they also have a critical perspective on the contingent physical and social circumstances that generate these risks. Thus, there is an ethically relevant difference between risks due to natural accident and risks due to social prejudice, for example. It is at least arguable that Sharon and Candace have adopted such a critical perspective. Consequently, it is not so obvious that they were wrong to select for deaf children. But even if they were, the explanation would not be exhausted by the consideration that they failed to maximise benefit impartially. Their failure would also consist in their inability to realise the essentially partial project of creating a flourishing family for themselves in their highly peculiar historical circumstances.

Candace and Sharon's defence does not generalise to all forms of disability. Human deafness is an unusual disability in several respects. First, it is a moderate disability compatible with living a very good life of its kind. Second, the existence of special forms of communication like sign language and lip reading enables deaf people to participate in a valuable form of social life without removing their particular disability. These facts have enabled a strong deaf culture to develop in many countries. Although similar considerations might apply to other disabilities like blindness, it does not apply to all disabilities. Thus, it is not so obvious that there is a distinctive Down's syndrome culture.[24] Nor is the absence of a disability culture confined to serious disabilities or impairments. There is no distinctive asthma culture either. Yet Down's syndrome and asthma also present pre-conception scenarios where choosing to conceive a child with the relevant condition is compatible with that child enjoying a good life of its kind. Whether it would always be wrong to deliberately conceive children with Down's syndrome or asthma will depend on the extent to which the reproductive context in question includes alternative compensatory features of the kind exhibited by the deaf case. Either way, it is an ethical distortion to assume that cases like these are all decidable exclusively by appeal to impartial principles like Q and the beneficence principle.

Non-identity and disabilities

The conventional view on disabilities has been clearly stated by Jonathan Glover: 'Consider the theoretical possibility of screening to ensure that only a disabled child would be conceived. This would surely be monstrous. And we think it would be monstrous because we do not believe it is just as good to be born with a disability.'[25] One explanation for the prevalence of the conventional view of disabilities is the widespread acceptance of what Parfit calls the *no difference view*.[26] On this view, it is wrong both to cause a disability in an existing person and to cause the existence of a disabled person, and for the same reason – namely, that the outcome is worse in terms of non-person-involving and impartial considerations. It makes no ethical difference that in one case the outcome is worse for an individual and the other not. If the arguments of the previous sections are sound, we have reasons to reject the no difference view. First, the reasons it would be wrong to cause disability in an existing person are not exhausted by the fact that the outcome would be worse in terms of non-person-involving considerations. Second, the reasons it would be wrong to cause disabled persons to exist are not exhausted by the consideration that the outcome would be worse impartially considered. Third, it is not obviously wrong to cause disabled persons to exist. Fourth, the reasons it might not be wrong to cause disabled persons to exist are not exhausted by the thought that the outcome would not be worse impartially considered. In so far as it rests on the no difference view, the conventional view of disabilities is based on an overly simplified picture of the ethics of reproduction.[27]

Several writers in the recent literature have rejected the no difference view, but continue to handle preconception scenarios impartially. Thus, McMahan claims that while causing disability through pre-natal choice is wrong because the effect is worse for an individual, to choose disability in a pre-conception scenario is wrong because the effect is worse non-personally and impartially considered. McMahan appears to endorse the conventional view of disabilities when he writes: 'What we need is an account that explains why it is objectionable to cause a disabled child to exist when it would be possible to cause a normal child to exist.'[28] Later in the same paper, when he discusses Kavka's account of disability as 'restricted life', McMahan considers how the conventional view might be doubted.[29] Yet he apparently ends up defending it:

> Assuming ... that the desire to have a child has a certain normative force ... it might be that the desire of a couple to have a child could be sufficient to outweigh the harm they would do to the child by causing it to exist with a restricted life. But this same desire would be insufficient to justify causing a child to exist with a restricted life when it would be possible to have a normal life instead ... There would have to be some *other* reason to justify doing what would cause a child with a restricted life to exist rather than a normal child. And in the ordinary circumstances of life it is doubtful that there could be a reason sufficiently strong to justify the harm to a child with a restricted life.[30]

Once we give up the no difference view, it is no longer clear that only the production of benefit impartially considered is ethically relevant in non-identity scenarios involving potential disability (perhaps this is what McMahan means by there having to be 'some *other* reason'). In both the superhuman and deaf cases, contextual considerations involving the partial concerns of the prospective parents pose a direct challenge to McMahan's implicit assumptions about what he calls 'the ordinary circumstances of life'. A parallel criticism can be made of Buchanan et al., who apply the conventional view of disabilities directly to the case of deafness:

> It may be possible to imagine a world in which a reasonable person, confronted with such a choice, would choose deafness, but this is not our world. To make such a choice reasonable for most people would require an enormous reallocation of social resources, indeed a radical restructuring of our modes of production and social institutions, in order to make it true that for most people who are deaf, the benefits of membership in the deaf community outweigh the limitations on opportunity that deafness brings.[31]

Candace and Sharon are not choosing for most people who are deaf. Nor is it clear that Buchanan et al.'s model of reasonably hypothetical choice is well suited or even coherently applied to preconception scenarios.[32] In any case, the relevant choice faced by a reasonable person in the deaf case is not whether to choose deafness in any arbitrary circumstances characteristic of our world, but whether to choose deafness in the particular context of a socio-economically privileged deaf community in early twenty-first-century California. Part of what draws Buchanan et al. to their negative conclusion about deafness is their theoretical concern with social justice and global considerations of equality of opportunity. These concerns might conceivably argue against the use of public funds to promote the incidence of disability in society at large on grounds of its foreseeable implications for the distribution of public resources. It does not follow that individual deaf parents are acting impermissibly by deliberately having deaf children and bringing them up by their own means. If the latter claim is entailed by the conventional view of disabilities, that view should be reconsidered.

One response to my criticism of Buchanan et al. is that it leaves public institutions like the legal system unable to decide when decisions in pre-conception scenarios require formal regulation or censure. While serious, this worry should not be exaggerated. The ethics of individual reproductive choice is not equivalent to its legality. Regardless of the merits of the deaf case, there are non-controversial cases of ethically unacceptable reproductive activity involving the disabled, such as the industrial production of severely retarded children for live experiments. Such activities are non-controversial candidates for formal regulation or censure. If the legality of selecting for disability through IVF were inevitably to cause non-controversially unacceptable reproductive activity, this would provide the basis for a slippery slope argument for its prohibition (analogous arguments are sometimes made about abortion and euthanasia). While this would be hard on some naturally infertile disabled couples, it could also be an ethical burden a state would have to impose in order to prevent even more serious ethical wrongdoing. On the issue of human reproduction, as in many other areas of public concern, ethics and the law are imperfect bedfellows.[33]

Conclusion

Agents are capable of making reasonable ethical evaluations in at least some non-identity scenarios. Such evaluations include, but are not exhausted by, non-person-involving and impartial considerations of benefit. Evaluations of non-identity scenarios also involve consideration of partial values peculiar to the individual case. While these partial values may partly define what counts as beneficial in a given scenario, they also partly determine the evaluative perspective from which these benefits can reasonably be evaluated. Like the notions of non-person-involving and impartial benefit, the interpretation of partial values is a source of conflict among reasonable people. It is therefore unsurprising that in the deaf case, for example, philosophers concerned with social justice have naturally adopted a different evaluative perspective than some deaf couples wanting to start a family. There is no immediate prospect of a resolution of these disputes. The evaluation of preconception scenarios is likely to remain a topic of ethical controversy.

While not obviously incoherent, the idea that one form of life is better than another impartially considered is a dangerous one. As often as not, the appeal to impartial benefit amounts to little more than the generalisation of one set of partial values to every conceivable case of a given type. In this way, a seemingly innocent commitment to impartial considerations like Q or the beneficence principle as constitutive of 'the morality of beneficence' can lead to a dehumanised picture of ethical thought, both in non-identity scenarios and elsewhere. Reflection on the different values at work in particular cases can sometimes produce a corrective to such tendencies.

Notes

1 D. Parfit (1984) *Reasons and Persons* (Oxford: Oxford University Press); and J. Woodward (1986) 'The Non-Identity Problem', *Ethics*, 96.
2 The distinctions cut across each other. Impartial person involving, impartial non-person-involving, partial person involving and partial non-person-involving considerations are all found in actual ethical discussion.
3 A. Buchanan, D. W. Brock, N. Daniels and D. Wikler (2000) *From Chance to Choice* (Cambridge: Cambridge University Press).
4 Given the ethical controversy about abortion, this label is potentially misleading. I shall nevertheless use it to minimise the use of philosophical jargon.
5 Parallel conclusions follow for the concept of harm. I shall take this as read in what follows.
6 J. Glover (2001) 'Future People, Disability and Screening', in J. Harris (ed.), *Bioethics* (Oxford: Oxford University Press).
7 Parfit, *Reasons and Persons*.
8 The beneficence principle is sometimes characterised in the literature as defining 'the morality of beneficence' (cf. Parfit, (1982) 'Future Generations: Further Problems', *Philosophy and Public Affairs* II, pp. 127–8; Woodward, 'The Non-Identity Problem'). While this term is officially intended to pick out a proper subset of ethical concerns, those who use it sometimes discuss

the non-identity problem as if the domain of beneficence and the domain of ethical concern were coextensive. While this appearance may be deceptive, it has arguably been a factor encouraging an overly narrow view of the values involved in non-identity scenarios.

9 Parfit, *Reasons and Persons*, p. 388.

10 Parfit, *Reasons and Persons*; Buchanan et al., *From Chance to Choice*; J. McMahan (2001), 'Wrongful Life: Paradoxes in the Morality of Causing People to Exist', in J. Harris, *Bioethics* (Oxford: Oxford University Press).

11 Parfit, *Reasons and Persons*, p. 360. McMahan discusses and rejects this principle under the label 'the impersonal comparative principle' in McMahan, 'Wrongful Life', p. 461. Buchanan et al. discuss and endorse what could be interpreted as a more elaborate version of the same principle, and which they call N: 'Individuals are morally required not to let any child or dependent person for whose welfare they are responsible experience serious suffering or limited opportunity or serious loss of happiness or good, if they can act so that, without affecting the number of persons who will exist and without imposing substantial burdens or costs or loss of benefits on themselves or others, no child or other dependent person for whose welfare they are responsible will experience serious suffering or limited opportunity or serious loss of happiness or good' (Buchanan et al., *From Chance to Choice Genetics & Justice*, p. 249).

12 Parfit, *Reasons and Persons*, p. 361.

13 Cf. Buchanan et al., *From Chance to Choice*; McMahan, 'Wrongful Life'.

14 Parfit, *Reasons and Persons*, p. 433; quoted in McMahan, 'Wrongful Life', p. 466.

15 Parfit, *Reasons and Persons*.

16 D. Parfit (1986) 'Comments', *Ethics*, 96, pp. 854–62.

17 Pets are peculiar animals in this respect because, unlike other animals such as farm animals, many are bred for no other purpose than the enjoyment they give to their owners.

18 This example is adapted from Case C in R. M. Adams (1972) 'Must God Create the Best?' *Philosophical Review* 81, pp. 329ff. Adams uses this case to argue against a principle he formulates as follows: 'It is wrong to bring into existence, knowingly, a being less excellent than one could have brought into existence' (p. 329). A similar case is criticized in D. Parfit (1982) 'Future Generations: Further Problems', *Philosophy and Public Affairs* 11, pp. 127–8.

19 Adams, 'Must God Create the Best?', p. 330. Adams calls this principle R. He goes on to claim that principle R is 'rooted in our obligation to God, as his creatures, to respect his purposes for human life' (pp. 330ff). While I am sympathetic to many of Adams' ethical conclusions, I am not committed to his theological premises.

20 The deaf case is an actual case from recent history (L. Mundy (2002) 'A World of Their Own', Washington Post, 31 March, p. W22; and H. Lillehammer (2003) 'Who Needs Bioethicists?', *Studies in the History and Philosophy of Biological and Biomedical Sciences*, pp. 131–44). The case provoked substantial controversy, although not necessarily for the right reasons. One reason for the controversy was the ethical status of the IVF technique. Another reason was the gender of Sharon and Candace, a lesbian couple from California. We can ignore these sources of controversy

here. The ethical problem of selecting for disabilities would remain had the biological mother procreated with a deaf male in the conventional way. An additional issue that affects the present argument is that the children of Sharon and Candace are educated in a specialised school paid for by public funds. It is a controversial issue whether public funding of specialised schools for deaf children would be a just requirement of non-discrimination or an unjust burden on others if deafness were easily avoidable. As this feature is not essential to the case either, I shall ignore it. Finally, I shall ignore the question whether in avoiding any form of treatment for the deafness of their children, Sharon and Candace thereby committed themselves to the claim that it would be right to produce a disability in an existing normal person. This conclusion would follow only if there were no ethically significant distinction between omitting to treat a disability on the one hand, and acting to produce a disability on the other. Some consequentialists reject this distinction; cf. J. Bennett (1994) *The Act Itself* (Oxford: Oxford University Press). Consequentialists are also the most likely defenders of impartial and non-person-involving considerations like Q and the beneficence principle. It is a moot question how consequentialists should account for the partial values discussed in the present chaper.

21 Cf. G. S. Kavka (1982) 'The Paradox of Future Individuals', *Philosophy and Public Affairs* 11, pp. 110–11, who explicitly extends the categorical imperative to 'forbid treating rational beings *or their creation* ... as a means only, rather than as ends in themselves' (my italics).

22 In this respect, Sharon and Candace's choice differs from Adams' Case A, where a normal couple 'become so interested in retarded children that they develop a strong desire to have a retarded child of their own – to love it, to help it realize its potentialities (such as they are) to the full, to see that it is as happy as it can be' (Adams, 'Must God Create the Best?', p. 326). Adams takes this example to be a paradigm of prospective parents doing 'something wrong', even though in doing wrong they do not wrong the resulting child. He appeals to his principle R to explain the source of their wrongdoing.

23 Cf. Mundy, 'A World of their Own'.

24 Cf. Buchanan et al., *From Chance to Choice*, pp. 281–5.

25 Glover (2001) 'Future People, Disability and Screening', p. 438.

26 Parfit, *Reasons and Persons*.

27 See note 8 above.

28 McMahan, 'Wrongful Life', p. 456.

29 Kavka, 'The Paradox of Future Individuals', pp. 93–112. A restricted life for Kavka is one 'deficient in one or more of the major respects that generally make human lives worth living' (p. 105). While Kavka concedes that a restricted life could be more worthwhile than an unrestricted life, he nevertheless seems to think it is wrong to knowingly produce restricted lives on the grounds that 'on average, restricted lives are less rewarding than unrestricted ones' (p. 105). Because of its reliance on such generalizations, Kavka's claim is vulnerable to the above criticisms of impartial principles like Q and the beneficence principle. Few prospective parents think of themselves as selecting for 'average' lives.

30   McMahan, 'Wrongful Life', p. 458.
31   Buchanan et al., *From Chance to Choice*, p. 283.
32   Cf. R. Kumar (2003) 'Who Can Be Wronged?' *Philosophy and Public Affairs* 31.
33   Thus, excellent drivers are not excused from speeding fines merely on the grounds that it is normally safe for them to drive faster than the speed limit.

2
'Designer Babies', Instrumentalisation and the Child's Right to an Open Future

Stephen Wilkinson

In recent years, there has been a great deal of public debate about the creation of so-called 'designer babies'. This somewhat sensationalist expression is in many respects unsatisfactory. It does however provide a useful and widely recognised shorthand for a range of reproductive practices and for a particular set of concerns relating to those practices. There are three main reasons why the 'designer babies' debate merits attention. The first is simply the enormous public interest in reproductive technology as evidenced by copious news reports and television and radio programmes on the subject. The second is that it raises a number of important theoretical issues. The third is that it has obvious relevance to the development of legislation and public policy. This chapter aims to explain what the 'designer babies' debate is and to critically assess two of the main arguments for prohibiting the creation of 'designer babies'.

What is the 'designer babies' debate?

Those who (without quotation marks) invoke the expression 'designer babies' usually do so in order to raise a certain kind of concern about certain kinds of practice. The practices are those which enable parents (or others) to make and implement choices about the nature of future children. Because of the present state of reproductive technology, most real-life examples of this are *selection* techniques. These include pre-natal testing of foetuses, followed by selective termination, pre-implantation testing of embryos followed by selective implantation, and pre-conception sorting of gametes followed by selective use. However, there is no reason why techniques which *modify* individual foetuses, embryos or gametes in accordance with parental preference cannot also be covered by the label 'designer babies'.

The moral concern conveyed by the expression 'designer babies' is best introduced by looking at other comparable uses of the word 'designer'. These occur most commonly in connection with clothing, jewellery, perfume and other personal accessories. Hence, people talk of 'designer clothes', 'designer watches', etc. When the phrase 'designer babies' is invoked, then, this is in order to suggest that parents' attitudes to their 'designer babies' are similar, and similar in morally problematic ways, to the sorts of attitude that people have towards their designer clothes and jewellery. In some cases, the comparison is supposed to be quite specific. For example, parents who select a particular embryo because it will come to have certain cosmetic features (say, blond hair and blue eyes) might be said to be choosing their child in just the same way that they choose they clothes – i.e. based mainly or solely on outward appearance. However, usually the comparisons made are not this direct and what is alluded to is a much more general notion such as treating the child 'as a commodity'. As such, 'designer babies' claims often go hand in hand with ones about commodification and about treating a child as a means to an end and such talk is rife in the 'designer babies' debate.

Some preliminaries and exclusions

Before getting down to the main business, a number of distinctions and preliminary points need to be made. The first is a distinction between issues of *personal morality and professional ethics*, on the one hand, and issues of *legal prohibition and regulation*, on the other. My primary concern is with the latter, with whether creating certain kinds of 'designer baby' ought to be permitted and, if so, under what circumstances and within what sort of regulatory environment (although, for reasons of space, I will not say much about particular regulatory environments). Having made this distinction, however, I should point out that in practice there is a great deal of overlap. For the very same arguments are often deployed in both the ethics debate and the one about law and policy, with many people believing that the reasons why making 'designer babies' is wrong are also reasons for prohibiting it.

The second distinction is between objecting to parents *being able to choose* a baby with (or without) characteristic x and objecting to *the particular techniques used to deliver this choice*. For example, say that the choice to have a baby with or without characteristic x can be delivered only by using pre-natal screening and selective termination. In this case, there will undoubtedly be some anti-abortionists who believe that, although there is nothing wrong in principle with choosing a baby with or without x, it is wrong to exercise this choice by using selective termination, because abortion is wrong. Sex selection is an interesting example to cite here. Until very recently, the only effective ways of picking your child's sex were selective abortion and, in the last few years, pre-implantation testing of embryos. Since these techniques involve the destruction of foetuses or embryos, most anti-abortionists generally opposed sex selection. However, it is now possible to use a technique called 'sperm sorting' to select the sex of your child.[1] So one might expect the anti-abortionists' opposition to sex selection (if done using sperm sorting) to diminish and, if it doesn't, this will suggest that their opposition to sex selection is driven by something other than just their desire to avoid the destruction of embryos or foetuses.

In the real-world public debate, this distinction between the wrong-ness of *making* certain choices and the wrongness of a particular *method of implementing* them is seldom observed, and so 'designer babies' issues get mixed up with wider, and often intractable, disputes about abortion and the rights of the embryo/foetus. None the less, for the pur-poses of the present philosophical discussion, I am going to assume (or stipulate) that the 'designer babies' debate is fundamentally about the rightness or wrongness of parents *being allowed to choose* certain traits. Hence, I will disregard worries about this or that *method* of delivering such choices, and especially arguments based on the alleged wrongness of destroying embryos. I am not, of course, suggesting that concerns about the status of the embryo are irrelevant. The point is rather that these objections don't count *specifically* against creating 'designer ba-bies', but instead against a very wide range of permitted practices, including IVF, abortion and some birth control techniques.

Furthermore, there are pragmatic reasons for not tackling 'head-on' questions about abortion and the status of the embryo in a paper about 'designer babies'. One is that these questions have already been addressed, and in many cases addressed excellently and in considerable detail, by philosophers and others and so it is best to avoid 'reinvent-ing the wheel' by rehearsing the abortion debate here.[2] Another is that, as I suggested above, many of the disagreements about abortion and related practices appear intractable. Hence, basing our 'designer babies' arguments on a particular view of the status of the embryo would prob-ably lead directly to argumentative deadlock. So it would be better for both sides if alternative arguments were explored, ones that *don't* rely on contentious assumptions (one way of the other) about the status of the embryo.

A third important distinction, one that often comes up in philosophical debates about reproductive ethics, is between identity-affecting and other choices. Identity-affecting decisions are those that affect not what life will be like for a fixed future population or person, but instead affect who will exist in the future. So where we have a choice between implanting Embryo A and thereby creating Person A and implanting Embryo B and thereby creating Person B, this choice is (arguably at least) an identity-affecting one – a decision to create one rather than another possible further person. Such choices are to be contrasted with non-identity-affecting decisions such as whether or not to subject a particular foetus to (non-lethal) pre-natal micro-surgery. Although this distinction is an important one to keep in mind throughout my discussion, I shall say very little about 'non-identity' issues here. There are two reasons for this. First, the non-identity problem is covered in considerable detail by Lillehammer in the first chapter in this volume. Second, most of the 'designer babies' arguments that I consider here apply to both identity-affecting and non-identity-affecting decisions, and are supposed to apply equally to modification and selection.

Fourthly, I should reiterate that this chapter considers only two of the many possible types of objections to 'designer babies': that it instrumentalises the child (or embryo)[3] and that it violates her right to an open future. Hence, my conclusions are necessarily limited to the success or otherwise of these argumentative approaches.

One of the arguments that I will not discuss in any detail does, however, deserve a particular mention at this point: the child welfare argument. Those who oppose the creation of 'designer babies' often cite the welfare of the children that will be created.[4] These claims are normally based on a widely held moral belief (one enshrined in English law) that, when making decisions about the use of reproductive technologies, we are under an obligation to take very seriously the welfare of any child created.[5] Some people go further and think it wrong deliberately (or even just knowingly) to create a person who will suffer from a serious disease or disability, or who will have severe psychological problems. If these moral principles are combined with empirical claims about 'designer babies' being harmed, either physically or emotionally, as a result of their unusual origins, then there is a powerful *prima facie* case against the use of at least some reproductive technologies.

The debate about the welfare of the child hinges on two questions.[6] First, there is an *empirical* question about whether or not 'designer babies' will be any worse off than 'normal' children. Those who think that will tend to cite either unknown risks to physical health, or psychological harm which may be caused when the child finds out that she has in some sense been 'designed' or 'artificially' created. Second, there is a more *philosophical* question: what exactly is the relevance of child welfare considerations? And, in particular, if it were established that 'designer babies' were (on average) less happy than other children, would this fact be sufficient to justify forbidding the creation of such babies? This is a more difficult issue than it might at first appear for, as Harris argues: 'To give the "highest priority to the welfare of the child to be born" is always to let that child come into existence, unless existence overall will be a burden rather than a benefit.'[7] Why, then, have I chosen to exclude child welfare arguments from the present discussion? This is mainly because, as Harris suggests, it is impossible properly to discuss child welfare arguments without engaging with the non-identity problem (which is covered elsewhere in this volume) – and also because engaging with the other non-philosophical aspects of the child welfare argument would require the marshalling of huge bodies of socio-psychological evidence (in so far as such evidence exists) thereby taking us far from the central moral-philosophical concerns of this chapter and this volume.[8]

Three examples: saviour siblings, sex selection and 'designer disability'

In order to help focus the discussion, I will describe three cases to which 'designer babies' discourse has been applied: saviour siblings, sex selection and 'designer disability'. The aim of this section is briefly to explain these cases. My main reasons for selecting these particular examples are: (a) that they are already real possibilities and ones that (at least in the UK) have been widely debated in relation to law and regulation, and (b) that each raises interesting and distinctive issues. Of course, this list of examples is by no means exhaustive and other interesting possibilities worth noting include selecting a child by reference to its cosmetic features, intelligence, or susceptibility to disease.

Saviour siblings

In July 2004, the British news media carried headlines such as 'Designer Baby Transplant Success'.[9] The stories were about Michelle and Jayson Whitaker's five-year-old son Charlie, who suffers from Diamond Blackfan anaemia (DBA), a rare form of anaemia where the bone marrow produces few, or no, red blood cells. Symptoms are similar to other forms of anaemia and include paleness, an irregular heartbeat and a heart murmur because of the increased work the heart needs to do to keep oxygen moving around the body. The disorder can lead to irritability, tiredness and fainting and requires intensive therapy, including painful day-long blood transfusions and daily injections.[10] One way of helping Charlie was to create a 'saviour sibling' using PGD (pre-implantation genetic diagnosis) and HLA tissue typing. This would involve creating a child in order to provide life-saving tissue for Charlie, although crucially this tissue would not come from the new child's body but rather from the umbilical cord after birth. The HFEA did not allow the Whitakers to use the procedure in the UK and so they were forced to go to the United States for treatment, treatment which appears to have been reasonably successful so far.[11]

The Whitakers were denied treatment in the UK because, at that time, the HFEA permitted the use of HLA tissue typing only in cases where the 'potential child' was itself at increased risk of suffering from a genetic disorder. The procedure would be allowed only where there was a dual purpose, preventing the birth of a child with a disorder *and* helping an existing child, not where helping the existing child was the only reason for intervening. Charlie's case did not meet this condition because the DBA from which he suffers was 'sporadic' rather than hereditary, so that that the likelihood of his parents having a second baby with the disease were no greater than those of the general population: 5–7 per million live births. As such, there is no reason to believe that the Whitakers' embryo would have the same defect. Interestingly, the HFEA recently dropped this policy and now allows HLA tissue typing in cases like the Whitakers', provided that certain other conditions are met.[12]

Sex selection

Sex selection has been in the news in recent times for a number of reasons. First, new techniques (such as sperm sorting and pre-implantation genetic selection) have made selecting a particular sex easier and more reliable. Second, the Human Fertilisation and Embryology Authority ran a consultation entitled *Sex Selection: Choice and Responsibility in Human Reproduction* in 2002.[13] Some commentators at that time suggested that this was a prelude to liberalisation, but the HFEA's eventual recommendation was that sex selection for 'non-medical' reasons should continue to be prohibited.[14] Third, a number of couples have hit the headlines because of their desire to sex-select. One of the most emotively and widely reported of these was the case of Alan and Louise Masterton. They had four sons and one daughter, Nicole, who tragically died in a bonfire accident.[15] The Mastertons wanted to use PGD to select a second daughter, but their request was turned down by the HFEA whose rules only permit sex selection in order to avoid sex-linked genetic disorders such as haemophilia, muscular dystrophy and cystic fibrosis. The Mastertons later travelled to Rome for treatment.[16]

'Designer disability'

Jeanette Winterson writing in the *Guardian* newspaper asks:

> In the long argument over designer babies, did anyone imagine that parents might prefer a designer disability? While we were all worrying about the bionic offspring of the super-rich, two deaf lesbians in America were going round the sperm-banks, trying to make a deaf baby. It sounds like the start of a bad joke, except that they have now managed it twice.[17]

In April 2002, Sharon Duchesneau and Candy McCullugh, an American lesbian couple, 'attracted fierce criticism by deliberately having a deaf baby', using a friend with five generations of deafness in his family as a sperm donor.[18] Duchesneau and McCullough have both been deaf since birth and 'are part of a growing movement in the US which sees deafness as a cultural identity, not as a disability.'[19] McCullough, the child's adoptive mother said:

> Some people look at it like 'Oh my gosh, you shouldn't have a child who has a disability!' but, you know, black people have harder lives. Why shouldn't parents be able to go ahead and pick a black donor if that's what they want? They should have that option. They can feel related to that culture, bonded with that culture.[20]

Savulescu, commenting on this case, points out both that some deaf couples already want to use pre-natal genetic testing (combined with selective termination) in order to achieve the same result, and that:

> These choices are not unique to deafness. Dwarves may wish to have a dwarf child. People with intellectual disability may wish to have a child like them. Couples of mixed race may wish to have a light skinned child (or a dark skinned child, if they are mindful of reducing the risk of skin cancer in countries like Australia).[21]

Furthermore, the HFEA and the Advisory Committee on Genetic Testing consulted on this and similar issues in 2000 as part their wider consultation on pre-implantation genetic diagnosis, noting that: 'The question ... arises whether it is right deliberately to cause a child to be born with a disability?'[22] Clearly, then, 'designer disability' is both a real possibility and one that provokes vigorous ethical debate, particularly because of the serious concerns that many people have about the welfare of children with 'designer disabilities'.

Arguments in favour of 'designing' babies

Before critically assessing two of the major arguments against the creation of 'designer babies', I want to sketch the main positive arguments in favour of permitting 'designer babies'. Such arguments generally focus either on the value of parental choice or on the practical benefits which are supposed to ensue.

The parental choice point is often couched in terms of 'procreative autonomy' or 'procreative liberty'. Dworkin defines this as a person's 'right to control their own role in procreation unless the state has a compelling reason for denying them that control'.[23] Framed in this way, the right to procreative autonomy doesn't directly entail any practical conclusions since, in any given policy area, we still need to know whether or not the state has 'a compelling reason' to restrict reproductive freedom. This perhaps explains why a great deal of the debate is concerned with critiquing the arguments *against* 'designer babies'. For while most of us can sign up to Dworkin's abstract principle, this still leaves most of the ethical work still to be done – in particular, the thorny question of which arguments for restricting choice are sound. Given this, the principle of procreative autonomy might appear trivial, since everything hangs on what counts as a 'compelling reason' for state interference. However, even if understood in this minimalist way, the principle is not completely vacuous, for it does at least set in place a *presumption* of non-interference and puts the *onus of proof* on prohibitionists and others who would seek to regulate or otherwise restrict people's reproductive behaviour.

Two main justifications lie behind the principle of procreative autonomy. The first says that procreative autonomy follows from a general commitment to autonomy, from the view that, in *all* areas of personal life, individuals have a right to control x unless the state has a compelling reason for denying them control over x. Hence, we get procreative autonomy simply by replacing 'x' with 'reproduction', just as we might get 'sports autonomy' by replacing 'x' with 'sporting activity', or 'nutritional autonomy' by replacing 'x' with 'eating'. Many people understandably find this justification unsatisfying because it fails to show why (or that) procreative autonomy is more important than sports autonomy and so fails to ground the *special* significance that many want to attach to it. For this reason, some people prefer the second type of justification which is well expressed by Robertson:

> Why should procreative liberty have moral or legal rights status? ... Quite simply, reproduction is an experience full of meaning and importance for the identity of an individual and her physical and social flourishing because it produces a new individual from her haploid chromosomes. If undesired, reproduction imposes great physical burdens on women, and social and psychological burdens on both women and men. If desired and frustrated, one loses the 'defence 'gainst Time's scythe' that 'increase' or replication of one's haploid genome provides, as well as the physical and social experiences of gestation, childrearing, and parenting of one's offspring. Those activities are highly valued because of their connection with reproduction and its role in human flourishing.[24]

On this view, the principle of procreative autonomy is grounded in the importance that most people assign to reproduction, in the practical implications that a decision to reproduce (or not) has for individuals, and in the close relationship between reproductive decision-making and human flourishing.

Let's turn now to look at the other kind of reason for permitting 'designer babies' – the practical benefits which are supposed to ensue. These might take a number of different forms and will depend on the particular kind of 'design' involved. We can usefully group the benefits into two broad categories: (A) the creation of people who are happier and/or healthier than those who would otherwise have been created through natural reproduction, and (B) benefits to particular third parties or society. (B), selecting on the basis of benefits to third parties, would include the selection of a saviour sibling (to benefit the saved sibling and its parents), sex selection on the basis of parental preference, and similarly the selection of cosmetic features solely in order to satisfy parental desires. As regards (A), some writers (notably Savulescu) have argued for a principle of procreative beneficence, according to which: 'couples (or single reproducers) should select the child, of the possible children they could have, who is expected to have the best life, or at least as good a life as the others, based on the relevant, available information'.[25]

On this view, using selection technologies to choose the 'best possible' children is not merely morally permissible, but morally obligatory. Practices in the procreative beneficence category include 'therapeutic' screening/selection for genetic disorders, health enhancement screening to select future people with unusually low risks of getting various diseases, and also screening/selection which has nothing (directly) to do with health such as choosing embryos on the basis of (likely) intelligence, beauty or athleticism (on the assumption that these features generally have a positive effect of quality of life).

I mentioned earlier that I am mainly interested in questions of legal prohibition and regulation, so we need to ask how the principle of procreative beneficence impacts on these. Using law to prevent people from acting in accordance with their moral duties may in exceptional circumstances be justified. None the less, the fact that x is morally obligatory is normally a powerful (though defeasible) reason for permitting x. So if the principle of procreative beneficence is true, then using reproductive technologies to choose the 'best possible' children ought (at least in the absence of strong countervailing reasons) to be permitted, since we ought to allow parents and others to act on their moral duties of procreative beneficence.

As with procreative autonomy, this is not the place to analyse the principle of procreative beneficence in detail. None the less, it is worth noting two things. First, procreative autonomy and procreative beneficence will sometimes pull in opposite directions. For example, there may be some 'designer disability' cases in which parents want to choose a child who is likely to have a lower quality of life than a child chosen at random. Similarly, in saviour sibling and non-therapeutic sex selection cases a child who is not the 'best possible' may be chosen in order to benefit a sibling or in order to satisfy its parents' desires. Second, many people think that procreative beneficence generates unpalatable results when applied to enhancement cases, such as Lillehammer's superhuman case (in this volume), in which Jack and Jill (prospective parents) have a choice between creating a 'normal healthy' child or one with 'superhuman intelligence' (as well as good health). Procreative beneficence entails that it would be wrong to select the 'normal healthy' child, a conclusion that many people find counterintuitive.

To conclude, then, there are two main reasons for positively supporting the creation of 'designer babies'. The first is a commitment to the principle of procreative autonomy. This can be seen either as merely an application of a general principle of respect for autonomy, or as a special principle justified by reference to the importance of human reproduction. Second, there are consequence-based arguments which look to the positive outcomes that creating 'designer babies' may generate. Some of these are based on broadly utilitarian calculations, simply weighing the total costs and benefits. Others are based on the principle of procreative beneficence. In the rest of the chapter I consider two objections to creating 'designer babies'.

Instrumentalisation

Perhaps the commonest objection to creating a 'designer baby' is that doing so (allegedly) involves instrumentalising the child, wrongfully treating it 'as a means to an end'. The recent debate in the UK about saviour siblings, for example, provoked the following comments: 'It is totally unethical. You are not creating a child for itself.'[26] And 'We would have very serious concerns that he is a commodity rather than a person.'[27] Such comments combine two worries: concerns about people having children for the wrong reasons, on the one hand, and concerns about the way in which the child will be treated by its parents, on the other. Thoughts of the second kind are predominantly concerns about the welfare of the child and so I will not discuss these here.

It seems that conceiving can be wrong if done for the wrong reasons. Conceiving a child in order later to eat it or torture it would be uncontentious, if extreme, supporting examples for this principle. The question then is: which reasons are the wrong reasons? One answer (closely associated with instrumentalisation arguments) is that a child should be wanted *for its own sake* and not for some other purpose. Boyle and Savulescu make the point as follows: 'The commonest objection ... is that it is wrong to bring children into existence "conditionally". This objection finds its philosophical foundation in Immanuel Kant's famous dictum, "Never use people as a means but always treat them as an end".'[28]

Straightforward versions of this argument are often defective in two ways. First, many of them rely on a misunderstanding of 'Kant's famous dictum'. This doesn't prohibit treating people as means, but rather prohibits treating them *merely* or *solely* as means. As Harris puts it: 'We all ... [treat people as means] perfectly innocuously much of the time. In medical contexts, anyone who receives a blood transfusion has used the blood donor as a means to their own ends ...'[29] On the

Kantian view, there is nothing objectionable about creating a baby as a 'means to an end' provided that it is also viewed and treated as a human being. So even if it can be established that a particular practice involves treating a baby as a means, this will not prove anything ethically unless it can also be established that it is being treated *merely* as a means.

The second difficulty for instrumentalisation arguments is that they must show why creating a 'designer baby' is *more* objectionable than having a child for other widely accepted reasons (or show that these widely accepted reasons shouldn't themselves be accepted). Some people, for example, have children in order to 'complete a family', to provide a playmate for an existing child, to improve a marriage, to delight prospective grandparents, or to provide an heir. So we must ask whether these behaviours are supposed to be condemned along with 'designer babies', since they all appear equally instrumental? If the answer is 'yes', this will make the instrumentalisation arguments under consideration appear less plausible, for they will seem excessively restrictive and to 'prove too much'. This is true *a fortiori* if we are looking not just at ethics but also at legal prohibition, for surely having a child in order to provide a playmate for an existing child oughtn't to be banned, even if this isn't a particularly good reason for conceiving – but if this isn't restricted then why prohibit creating a 'designer baby', if both are equally instrumental?

Instrumentalisation arguments against creating a 'designer baby' face two fundamental problems. First, there are difficult basic questions about which things we are required to treat ends-in-themselves – questions which are very closely tied up with the moral status of embryos and foetuses. Second, as was just noted, treating someone instrumentally does not on its own entail a breach of 'Kant's famous dictum' since it is possible to treat persons both instrumentally and as ends-in-themselves. So proponents of the instrumentalisation argument are required to show *why* particular ways of treating one's (future) offspring instrumentally are incompatible with respecting them as ends-in-themselves. Expressed as questions, these two fundamental problems can be summed up as:

(A) Are we required to respect gametes, or embryos, or foetuses (depending on the stage at which selection takes place) as ends-in-themselves?

(B) If we are, which particular ways of treating them instrumentally are incompatible with respecting them as ends?

I will not attempt to answer (A) directly here, but what I can say with some confidence is that *if* the answer to (A) is 'yes' with respect to embryos, then a very large number of reproduction and selection techniques will be ruled out, regardless of whether or not they are used to create a 'designer baby'. This is because if we are required (for example) to respect an embryo as an end-in-itself, then that must entail a fairly strong *prima facie* duty not to destroy it. But, since IVF typically involves creating more embryos than are required for implantation and subsequently destroying 'spares', even regular IVF treatments must be ruled out.[30] What follows from this is that the instrumentalisation objection to creating a 'designer baby' collapses into a much more general 'moral conservative' objection to IVF of all kinds. For the instrumentalisation objection can only work (at least in its direct form) if there is something approaching a fairly strong *prima facie* duty not to destroy embryos. For our purposes, this means that the (direct form of the) instrumentalisation argument can be disregarded, for I said at the outset that the rightness or wrongness of parents being able to choose certain traits was my focus, not worries about this or that method of delivering such choices – and, in particular, arguments based on the alleged wrongness of destroying embryos. So since the instrumentalisation objection seems to rely on ascribing to embryos something rather like personhood (status as an 'end'), the direct version of this argument must be disregarded for the present. For what we are looking for here is an argument, or set of arguments, which can show creating 'designer babies' to be wrong even if other practices such as abortion and IVF are not. All of these considerations incidentally apply to gametic selection technologies *a fortiori* since in order to get the direct instrumentalisation argument to count against gametic selection, we would have to assign the status of 'ends' to individual ova and sperm.

Perhaps the only feasible version of the instrumentalisation argument therefore is an indirect one. What this says is that, although creating a child for instrumental reasons isn't intrinsically wrong, it may (in certain circumstances) be extrinsically wrong either because of effects on parents' attitudes to the child once it is born, or because being willing to create a child for instrumental reasons is symptomatic of inappropriate attitudes towards children that the parents already have. For the most part, this argument collapses into the child welfare argument, which I am not considering for reasons given earlier. It is, however, worth noting that there is a possible version of the argument which doesn't concern child welfare as such, but rather autonomy and instrumentalisation.

This argument is clearly articulated by Dena Davies, who during a discussion of whether deliberating creating a deaf child is a 'moral harm', states: 'creating a child who will be forced irreversibly into the parents' notion of "the good life" violates the Kantian principle of treating each person as an end in herself and never as a means only'.[31] Davies rightly acknowledges that there is nothing necessarily wrong with creating a child for instrumental or selfish reasons:

> We choose to have children for myriad reasons, but before the child is conceived those reasons can only be self-regarding. The child is a means to our ends: a certain kind of joy and pride, continuing the family name, fulfilling religious or societal expectations, and so on.[32]

But:

> Parental practices that close exits virtually forever are insufficiently attentive to the child as end in herself. By closing off the child's right to an open future, they define the child as an entity who exists to fulfil parental hopes and dreams, not her own.

So interestingly, for Davies, instrumentalisation arguments and 'right to an open future' arguments are very closely connected because what makes denying a child an open future wrong is that it is a way of failing to respect the child as an end in itself.

The child's right to an open future

This leads us naturally to look at the view that creating 'designer babies' is wrong because doing so violates what Feinberg calls 'the child's right to an open future'.[33] In order to assess this claim we need to ask both what the 'right to an open future' is, and which prenatal choices (if any) would constitute violations of this right.

What is the child's right to an open future?

According to Feinberg, the 'child's right to an open future' is a convenient shorthand for a set of rights with a certain form, a 'vague formula' that 'simply describes the form of the particular rights in question ... not their specific content'.[34] These are 'rights-in-trust', which he explains as follows:

'rights-in-trust' look like adult autonomy rights ... except that the child cannot very well exercise his free choice until later when he is more fully formed and capable. When sophisticated autonomy rights are attributed to children who are clearly not yet capable of exercising them, their names refer to rights that are to be *saved* for the child until he is an adult, but which can be violated 'in advance', so to speak, before the child is even in a position to exercise them. The violating conduct guarantees *now* that when the child is an autonomous adult, certain key options will already be closed to him. His right while he is still a child is to have these future options kept open until he is a fully formed, self-determining adult capable of deciding among them.[35]

The idea of *rights-in-trust* is a useful way of conceptualising certain moral issues. Take, for example, a case in which a mother decides that her young daughter would be better off not having children and so asks a doctor to sterilise her. Most of us, I imagine, would think that this sterilisation ought not to be permitted. This is partly for welfare reasons; the child will be distressed either now or later and may miss out on the opportunity of having positive parenting experiences. But even leaving welfare considerations aside (or even if we think the daughter really would be better off without children) moral objections to the sterilisation remain. Foremost among these is the child's right to an open future with respect to procreation. When she becomes an adult woman she will have a right to procreative autonomy. She doesn't presently have this right; rather, it is a right-in-trust. But if we sterilise her now, then we will (as Feinberg puts it) be 'violating in advance' her right to procreative autonomy before she has a chance to exercise it.[36]

This is a relatively (although not entirely) uncontentious application of the child's right to an open future. More controversial are cases in which parents (and others) seek to influence children's beliefs and values. For example, Feinberg discusses some American legal cases in which Amish communities have tried to keep their children out of state-accredited schools. This is thought to be incompatible with the goal of Amish education which is:

to prepare the young for a life of industry and piety by transmitting to them the unchanged farming and household methods of their ancestors and a thorough distrust of modern techniques and styles that can only make life more complicated, soften character, and corrupt with 'worldliness'.[37]

For several reasons, cases like this are more perplexing. One is that – if we take seriously the Amish claim that state education is, in some respects, polluting of young minds – it looks as if future options will be foreclosed whichever educational choice is made. For if corrupted (by state schooling) as a child one cannot, as an adult, simply *choose* to become uncorrupted; it may be irreversible. A second is that the extent to which options are foreclosed by education is uncertain, since the effects of exposing children to different kinds of schooling is relatively unpredictable. While, third, it should be noted that many widely accepted parenting and schooling practices are themselves designed to foreclose options – for example, they attempt to stop people from deciding to become criminals.³⁸ So one might ask whether 'main-stream' education is so fundamentally different from Amish education in this respect. For both are, as it were, biased in favour of generating adults who will make certain choices – the difference being over which choices are favoured.

Do embryos, foetuses and gametes have rights to an open future?

So far we've been looking at children and their rights to an open future. But even if we think that children have rights to a open future in certain areas of life, it doesn't obviously follow from this that embryos, gametes and foetuses have such rights. And, if they don't, then the idea of a right to an open future will be inapplicable to the 'designer babies' debate since these are the entities about which parents and reproductive scientists make decisions.

One obvious objection to ascribing a right to an open future to (say) an embryo is that this seems to suggest that the embryo has a right to life. For how can a being have a right to an open future without having a right to a future? As I made clear earlier, I am not going to tackle head-on the question of whether embryos have rights to life. However, what I can say is that *if* it were true that an embryo could have a right to an open future only if it also had a right to life, then this would be a serious blow to the 'right to an open future' argument against creating 'designer babies', since it would (like some other arguments) collapse into a much more general moral conservative argument against abortion and against a very wide range of reproductive practices. In other words, it would not be of any use as a *specifically* 'anti designer babies' argument.

There do, however, seem to be reasons for thinking that a right to an open future does not entail a right to a future, or needn't do so if understood in a certain way. We need at this point to distinguish strong from weak versions of the right to an open future. On the strong version, the right to an open future is a conjunction of two other rights: a right to a future (a right to life) and a right, *if you exist* (if you have a future), not to have certain options foreclosed. On the weak understanding, the right to an open future is just the second of these, a conditional right (conditional upon existence at the relevant point in the future). So, on the weak understanding, the form of the right is like that of the right to paint your house whatever colour you choose. This right doesn't contain or entail the right to a house. Rather, *if you have a house*, you can paint it whatever colour you like. For our purposes the weak understanding is to be preferred because, as I argued above, using the stronger version would mean that the argument collapses into a much more general 'conservative' argument. Of course, this doesn't mean that embryos (or children for that matter) don't have a right to life. It just means that, if they have such a right, this is not entailed by, or part of, the right to an open future.

That deals with the suggestion that the right to an open future entails a right to life. None the less, one might still be sceptical about the embryo having a right to an open future on the grounds that this would require it to have rights. This is a worry because many people would not want to ascribe rights to embryos (and *a fortiori* not to gametes) and so the 'right to an open future' argument could again be weakened by relying on an extremely contentious assumption. At this point I would concede that framing the argument in terms of rights is not helpful and would suggest that the idea lying behind the right to an open future can be better captured in a non-rights-based principle such as: *it is wrong (prima facie) to create, or help to create, a person whose future is insufficiently open in relevant respects, or to alter a person's life in ways which cause his/her future to be insufficiently open in relevant respects.* This principle seems to do everything that we want the 'right to an open future' to do, but avoids the metaphysical conundrums associated with rights talk (such as the moral status of the putative rights bearer).

Earlier, we looked at a case in which a mother decides that her young daughter would be better off not having children and so asks a doctor to sterilise her. I suggested that the idea of a child's right to an open future captures many people's intuitions about this case. I would also suggest that the non-rights-based principle outlined above, which we can call the Open Future Principle, captures our intuitions about this case equally well – and moreover that it is to be preferred in some other cases. Imagine, for example, a parallel situation in which a mother decides that her future child would be better off not having children and so asks a doctor to perform a pre-natal sterilisation (or some other pre-natal intervention with the same effects). In the original child sterilisation case there are, as we saw, various grounds for objecting. Some of these have to do with the child's (present) welfare and rights, but others are purely future-directed, to do (for example) with wrongfully limiting a future adult's procreative autonomy. These future-directed considerations must presumably apply equally to the original child sterilisation case and to the pre-natal intervention case, since what is at issue is not the present status or rights of the foetus/ child but rather the wrongness of limiting a future adult's procreative autonomy. The Open Future Principle captures this symmetry perfectly by focusing on the wrongness of creating a life that is insufficiently open and/or of causing a life to be insufficiently open – *regardless of the moral status of the entity on which the person now acts*. For this reason, the Open Future Principle is to be preferred to the idea of a *right* to an open future, because the latter (it seems) requires us to attribute rights (or at least 'rights-in-trust') to the entities on which we act in order for open future considerations to apply. Of course, looking at the cases in hand, some people do think that embryos are rights-bearers and for them this will not be a problem. But it will certainly be a problem for (almost?) everyone when we turn to possible interventions affecting gametes. For example, what if we could modify sperm in ways which drastically reduced a future adult's autonomy? In such situations, the Open Future Principle could apply and (depending on how 'insufficiently open in relevant respects' is interpreted) be used in an argument against such sperm modifications. But those who want to invoke the right to an open future would be forced instead to choose between abandoning open future arguments altogether or (implausibly) ascribing rights (or rights-in-trust) to sperm.

Applying the Open Future Principle to the 'designer babies' debate

Let's turn now to the more practical question of how the Open Future Principle applies to our three case studies: saviour siblings, sex selection and 'designer disability'. This means asking whether any of the following practices involve creating a person whose future is insufficiently open in relevant respects:

(A) selecting a future child that will be a tissue match for a sick sibling;

(B) selecting a male/female future child (for 'non-therapeutic' reasons);

(C) selecting a future child that will be deaf.

Sex selection appears not to violate the Open Future Principle directly because being biologically female or male is not generally a choice that autonomous adults have and so a sex-selected person would not be deprived of any choices. His or her future is no less open than anyone else's in relation to sex. Of course, it is possible for people to choose to change their sex surgically, etc., but this choice is as available to a sex-selected person as it is to anyone else.[39]

Turning now to the saviour siblings issue, it is important first to distinguish between different stages at which donation might take place. Recent discussions in the English courts and in the bioethics literature have been predominantly about the easiest case, one in which the only donation envisaged is of umbilical cord blood.[40] This is the easiest (i.e. least ethically problematic) case because the use of cord blood does not involve doing anything at all to the child created. More difficult would be cases in which donation by the saviour sibling during childhood or adulthood was envisaged, since direct physical injury and pain may be involved and there may also be consent issues.

Taking the easiest case first, saviour sibling selection of this kind does not directly violate the Open Future Principle because being a tissue match or not for one's siblings is not generally a choice that autonomous adults have and so an adult saviour sibling would not be deprived of this choice. Open future issues may, however, be thought to arise when adult donation is envisaged (and similarly where childhood donation is planned, although for reasons of brevity and simplicity I will restrict the present discussion to adults). Leaving aside the possibility of forced tissue removal, which would itself be unethical for other reasons, the worrying scenario is one in which an adult is expected by her family to donate life-saving tissue to a sibling. For, while donation may be formally optional in these circumstances, the weight of family expectation and thoughts like 'we brought you into existence to save your sibling, and you wouldn't exist if it weren't for her' may mean that the (prospective) saviour sibling feels she has little or no choice in the matter – particularly if the result of her refusing to donate would be the death of her sibling. Hence, it might be argued, her future is closed in so far as she is, in effect, compelled to donate tissue to a needy sibling.

The concern about people being pressured into donating is certainly a very proper one, but it is not clear that it can underpin a strong open future argument against selecting saviour siblings. It is worth noting initially that the position of the (prospective) saviour sibling is not fundamentally different from that of many other (prospective) living related donors, since they too may be vulnerable to family pressure (although admittedly not to the claim that 'we brought you into existence to save your sibling').[41] In both cases, the prospective donor has a choice – to donate or not – but the voluntariness of that choice and of any subsequent consent is called into question by the family's (intentional or otherwise) behaviour. But similarly, in both cases, any healthcare professionals involved in the tissue transfer have a duty to ensure that the consent given by the donor is valid, and in particular that it is genuinely voluntary – and we should assume, for our present purposes that they act on this duty. I am not suggesting that they always would act on this duty, but if we don't assume this, then we are muddying the waters by adding an independent wrong to the situation, one which isn't directly related to saviour sibling selection.

Given this assumption, would the saviour sibling's future be made insufficiently open by the possibility (or expectation) of adult donation? We can go some way towards answering this by comparing the (prospective) saviour sibling to two other possible people. The first is the 'natural' (prospective) saviour sibling who is not selected but just happens to be a tissue match. This person's degree of choice is essentially the same as the selected saviour sibling's (leaving aside the difference noted above, that the latter may receive more family pressure). The second is the non-tissue-matching sibling; if anything, her level of choice is *lower* than that of the saviour sibling, since she can't donate even if she wants to. Indeed, even in a situation where the saviour sibling was effectively forced into donating, she would still have the same level of choice as the non-tissue-matching sibling, since while the former would have to donate, the latter would have not to donate; neither has a choice. Now there are, of course, important differences between these two. In particular, being a donor (unlike being a non-donor) often involves physical risk and pain. However, considerations such as these are really part of *welfare* arguments, not *open future* arguments. I conclude, then, that the open future argument against selecting saviour siblings is unlikely to work. Saviour siblings' futures do not seem to be less open than those of relevant comparators, particularly if healthcare professionals and others ensure that any consents given are informed and voluntary.

Finally, is deliberately creating a deaf child contrary to the Open Future Principle? Davis is one bioethicist who (invoking the child's right to an open future) asserts forcefully that it does:

If deafness is a disability that substantially narrows a child's career, marriage, and cultural options, then deliberately creating a deaf child counts as a moral harm, because it so dramatically curtails the child's right to an open future. If deafness is a culture, as deaf activists assert, then deliberating creating a deaf child who will have only limited options to move outside of that culture also counts as a moral harm ... A decision made before a child is born that confines her forever to a narrow group of people and a limited choice of careers so violates the child's right to an open future that no genetic counsellor should acquiesce to it.[42]

Because of the complex empirical issues involved, it isn't possible to say definitively whether Davis is right about the particular case of deafness. But, on the face of it, her argument seems sound. For deafness does appear significantly to reduce the number of options available to people (elsewhere in *Genetic Dilemmas*, Davis cites various bits of evidence to back this up).

That said, there are a number of moves that someone who wanted to resist her conclusion could make. One is to question the empirical premise and argue that, while deafness closes down some options, it opens up others; for example, good relations with other members of the deaf community may be possible only for a person who is herself deaf. This line of argument, however, seems implausible since, even if it is true (which it probably is) that being deaf opens up options, the number of options it closes down is vast. Even if we confine ourselves to cultural and social activities, the hearing community far outnumbers the deaf one, and so one would expect many more socio-cultural options to be provided by it (although there is an interesting problem about how we individuate options).[43] So, given a choice between the two communities, we can expect choosing the deaf one to lead to a net reduction in the number of options available. Furthermore, it may not be necessary to be deaf in order to access some of the goods that membership of the deaf community provides. Savulescu makes the point as follows:

Hearing children of deaf parents can learn to sign, just as children of English parents can learn to speak Chinese as well as English. It is better to speak two languages rather than one, to understand two cultures rather and one. (It would be disabling for children of English parents living in China if their children spoke only English, even though it might be easier for their parents to communicate with them.)[44]

A second, more promising, rejoinder to Davis is to question just how far the child's right to an open future might take us if consistently applied. Take, for example, a small island community in the middle of the Pacific with its own language and culture, and let's assume for the sake of argument that this culture is 'limited' (whatever that means). Assuming that they are able and can afford to do so, are parents on the island obliged to expose their children to external cultural influences in order to give them a sufficiently open future? Similarly, if I had children in England, ought I to make them learn Chinese and Spanish in order to open up additional cultural options to them? I raise these examples in order to suggest that choosing whether or not to have a deaf child may not be *fundamentally* different from the routine choices that all parents have to make.

Similarly, it could be argued that the belief in a right to an open future is vulnerable to an objection levelled earlier at principle of procreative beneficence – that it makes enhancement-selection obligatory. Think back to Lillehammer's superhuman case in which Jack and Jill have a choice between creating a normal child or one with superhuman intelligence. Why not argue that, because of the child's right to an open future, they are obliged to select the superhumanly intelligent child, for surely such a child would have many more options available to it, a more open future, than a merely normal child? The obvious response to this is to say that we aren't obliged to *maximise* the openness of children's futures, but rather to ensure that they have a sufficient number of options (in certain respects). Maybe this is enough to distinguish enhancement/normality choices from deafness/normality choices, but we can at least see that there are some tough questions about the level at which to place the 'sufficient openness' threshold, and it is by no means obvious that this threshold should be normality, or that it should be higher than the number of options enjoyed by a typical deaf person.

Conclusion

This chapter provides a critical overview of two arguments for prohibiting the creation of (some kinds of) 'designer baby': the view that this practice involves instrumentalising the child and wrongfully treating it merely as a means to an end, and the view that it violates the child's right to an open future.

Instrumentalisation arguments against permitting the creation of 'designer babies' face a number of problems. One is that, in general, there seems to be nothing wrong with conceiving a child for instrumental or selfish reasons provided that, once it exists, it is treated appropriately: for example, wanting an heir to maintain the family estate seems not to be wrong in itself (and *a fortiori* ought not to be banned). A second is that, when considering reproductive choices, it is not clear that the entities in question (gametes, embryos, foetuses, etc.) really are ends-in-themselves – or, if they are, then instrumentalisation objections against 'designer babies' turn out just to be *general* 'moral conservative' arguments and thus fall outside this particular debate.

Faced with these problems, proponents of the instrumentalisation argument may fall back on an *indirect* version. What this says is that, although creating a child for instrumental reasons isn't intrinsically wrong, it may be extrinsically wrong either because of its effects on parents' attitudes to the child once it is born, or because being willing to create a child for instrumental reasons is symptomatic of inappropriate attitudes towards children that the parents already have. For the most part, this argument collapses into the child welfare argument, which is outside the remit of this chapter. However, there is a possible version of this argument (suggested by Davies) which relies upon the idea of the child's right to an open future.

While the child's right to an open future seems an attractive idea in some cases, it too is problematic and I have argued that an Open Future Principle is to be preferred. The ways in which such a principle might be applied to our three case studies (sex selection, saviour siblings and deliberating creating a deaf child) were also explored. I concluded that in the first two cases, it would be hard to construct a plausible open future argument, since the options available to any children created would not be constrained in ways that were both relevant and sufficiently serious. As regards selecting a deaf future child, there does seem to be a more plausible open future argument available here, just because not being able to hear is a constraint on options, although it should be noted that it may also open up some additional options. However, doubts remain about some fundamental aspects of the Open Future Principle – in particular, about what having a future that is 'insufficiently open' means, and about how we might non-arbitrarily set a threshold for a sufficiently open future.

Notes

1 HFEA (2003) *Sex Selection: Options for Regulation* [a report on the HFEA's 2002–3 review of sex selection including a discussion of legislative and regulatory options], available online at www.hfea.gov.uk.
2 To mention just a few: J. Finnis (1973) 'The Rights and Wrongs of Abortion', *Philosophy and Public Affairs*, 2.2, pp. 117–45; R. Hursthouse (1991) 'Virtue

Theory and Abortion', *Philosophy and Public Affairs*, 20(3), pp. 223–46; W. Quinn (1984) 'Abortion: Identity and Loss', *Philosophy and Public Affairs*, 13(1), pp. 24–54; J. Thomson (1971) 'A Defense of Abortion', *Philosophy and Public Affairs*, 1(1), pp. 47–66; M. Tooley (1972) 'Abortion and Infanticide', *Philosophy and Public Affairs*, 2(1), pp. 37–65; R. Wertheimer (1971) 'Understanding the Abortion Argument', *Philosophy and Public Affairs*, 1(1), pp. 67–95.

3 On this point, cf. David Oderberg, chapter 5 in this volume.

4 Emily Jackson (2001) *Regulating Reproduction* (Oxford: Hart), p. 173.

5 Human Fertilisation and Embryology Act 1990, s. 13(5). The importance of child welfare is a general principle of family law, most notably child welfare must be the paramount concern in all decisions where the upbringing or administration of a child's property is before the court, see the Children Act 1989, s. 1(1).

6 Stephen Wilkinson (2003) *Bodies for Sale: Ethics and Exploitation in the Human Body Trade* (London, Routledge); Sally Sheldon and Stephen Wilkinson (2001) 'Termination of Pregnancy for Reason of Foetal Disability: Are There Grounds for a Special Exception in Law?', *Medical Law Review*, 9, pp. 85–109.

7 J. Harris (2000) 'The Welfare of the Child', *Health Care Analysis*, 8, p. 33.

8 See also A. Campbell (2000) 'Surrogacy, Rights and Duties: A Partial Commentary', *Health Care Analysis*, 8, pp. 38–9; E. Jackson (2002) 'Conception and the Irrelevance of the Welfare Principle', *Modern Law Review*, 65, pp. 176–203; L. Purdy (1989) 'Surrogate Mothering: Exploitation or Empowerment?', *Bioethics*, 3; Wilkinson (2003) *Bodies for Sale*, pp. 149–59.

9 BBC News Online, 'Designer Baby Transplant Success', 27 July 2004, http://news.bbc.co.uk/1/hi/health/3930927.stm.

10 S. Sheldon and S. Wilkinson (2004) 'Hashmi and Whitaker: An Unjustifiable and Misguided Distinction?, *Medical Law Review*, 12, pp. 137–63.

11 BBC, 'Designer Baby Transplant Success'; B. Marsh (2004) 'Designer Baby Saving Brother's Life', *Daily Mail*, 27 July, www.dailymail.co.uk.

12 HFEA (2004) *HFEA Agrees to Extend Policy on Tissue Typing*, press release, 21 July, www.hfea.gov.uk/PressOffice/Archive/1090427358.

13 HFEA (2002), *Sex Selection: Choice and Responsibility in Human Reproduction*, www.hfea.gov.uk; Soren Hom (2004) 'Like a Frog in Boiling Water: the Public, the HFEA, and Sex Selection', *Health Care Analysis* 12(1), pp. 27–39.

14 HFEA (2003), *Sex Selection: options for regulation*, www.hfea.gov.uk.

15 BBC News, *Couple Fight for Baby Girl*, 4 October 2000, http://news.bbc.co.uk/1/hi/scotland/955251.stm.

16 In another reported case, a woman with four sons who wanted a daughter travelled to a Spanish clinic for sex selection using PGD. BBC News, *Mother Chooses Sex of Next Child*, 7 January 2003, http://news.bbc.co.uk/1/hi/england/2632827.stm.

17 Jeanette Winterson (2002) 'How Would We Feel if Blind Women Claimed the Right to a Blind Baby?', *Guardian*, 9 April, www.guardian.co.uk/Archive/Article/0,4273,4390038,00.html.

18 David Teather (2002) 'Lesbian Couple Have Deaf Baby by Choice', *Guardian*, 8 April, www.guardian.co.uk/international/story/0,3604,680616,00.html.

19   BBC News, 'Couple "Choose" to Have Deaf Baby', 8 April 2002, http://news. bbc.co.uk/1/hi/health/1916462.stm.
20   Teather, 'Lesbian Couple Have Deaf Baby by Choice'. See also M. Spiggs (2002) 'Lesbian Couple Create a Child who is Deaf Like Them', *Journal of Medical Ethics*, 28 (October), p. 283.
21   Julian Savulescu (2002) 'Deaf Lesbians, "Designer Disability" and the Future of Medicine', *BMJ*, 325 (10 May), p. 771.
22   HFEA and ACGT (2000) *Consultation Document on Preimplantation Genetic Diagnosis*, pp. 11–12, www.hfea.gov.uk/Downloads/Consultations/PGD/ pgdpaper.pdf. They continue: 'If a pregnant woman was found to be carrying a foetus affected by a disorder, it would not be considered appropriate to insist that she has a termination. The choice of whether to continue with the pregnancy in these circumstances would largely rest with the woman. However, in the case of PGD, because a pregnancy has not been established the nature of the choice to be made is different in that it involves a decision to begin a pregnancy knowing that a child would be born with a genetic disorder. The situation is further complicated because, by law, the clinician responsible for the treatment involving the use of PGD must consider, prior to treatment, the welfare of any child that might be born.'
23   Ronald Dworkin (1993) *Life's Dominion: an Argument about Abortion and Euthanasia* (London: HarperCollins), p. 148.
24   John A. Robertson (2003) 'Procreative Liberty in the Age of Genomics', *American Journal of Law and Medicine*, 29, pp. 439–87, at p. 450.
25   Julian Savulescu (2001) 'Procreative Beneficence', *Bioethics*, 15, pp. 413–26, at p. 415.
26   Josephine Quintavalle, quoted in BBC News, 'Doctor Plans "Designer Baby" Clinic', 11 December 2001, http://news.bbc.co.uk/1/hi/health/1702854.stm. Quintavalle is a leading member of the group *Comment on Reproductive Ethics*, which brought the judicial review action described above.
27   Vivienne Nathanson quoted in BBC News, 'Baby Created to Save Older Sister', 4 October 2000, http://news.bbc.co.uk/1/hi/health/1702854.stm.
28   Robert Boyle and Julian Savulescu (2001) 'Ethics of Using Preimplantation Genetic Diagnosis to Select a Stem Cell Donor for an Existing Person', *BMJ*, 323, pp. 1240–3, at p. 1241.
29   J. Harris (1985) *The Value of Life* (London: Routledge), p. 143.
30   Spare embryos may alternatively be donated to other women receiving treatment, or used in research (although the latter would itself lead eventually to destruction).
31   Dena Davis (1997) 'Genetic Dilemmas and the Child's Right to an Open Future', *Hasting Center Report*, 27(2) (March/April), pp. 7, 8.
32   Davis, 'Genetic Dilemmas'.
33   Joel Feinberg (1992) 'The Child's Right to an Open Future', in Joel Feinberg, *Freedom and Fulfilment* (Princeton: Princeton University Press), pp. 76–97.
34   Feinberg, *Freedom and Fulfilment*, p. 77.
35   Feinberg, *Freedom and Fulfilment*, pp. 76–7.
36   Davis, 'Genetic Dilemmas'.
37   Feinberg, *Freedom and Fulfilment*, p. 81.
38   Claudia Mills (2003) 'The Child's Right to an Open Future?', *Journal of Social Philosophy*, 34(4), pp. 499–509, at p. 500.

39   Another empirical argument is as follows. Sex selection (at least sometimes) indirectly violates the Open Future Principle by encouraging parents to subject their children to excessive gender stereotyping. The thought is that, having gone to the trouble of selecting a girl, for example, rather than just accepting whichever sex comes along, parents will feel committed to ensuring that their girl is not in any way boyish or masculine and has as many distinctively 'feminine' characteristics as possible. This in turn may mean that the girl's future is insufficiently open in a variety of ways. For example, she may be subjected to something akin to brainwashing in relation to how girls and women should behave, or may be deprived of education in 'masculine' areas such as engineering, science and vigorous sports.

In response to this, we should certainly concede right away that subjecting children to indoctrination and depriving them of education *can* render their futures insufficiently open. However, whether we thereby have a convincing argument against sex selection is much less clear. For the obvious flaw in the argument is its reliance on an empirical premise which, as far as I am aware, remains unproven: the claim that sex-selected children are more likely to be the victims of sexist indoctrination than others. Furthermore, it should be noted that many parents engage in sexist and other forms of indoctrination and so the argument for banning sex selection on anti-indoctrination grounds applies equally to many other parenting practices. Thus, it would seem arbitrary and unfair to single out sex selection for restriction.

40   Sheldon and Wilkinson, 'Hashmi and Whitaker: an Unjustifiable and Misguided Distinction?', pp. 137–63.

41   J. Harvey (1990) 'Paying Organ Donors', *Journal of Medical Ethics*, 16, p. 119.

42   Dena Davies (2001) *Genetic Dilemmas* (New York: Routledge), pp. 64–5.

43   Mills, 'The Child's Right to an Open Future?', p. 500.

44   Julian Savulescu, 'Deaf Lesbians, "Designer Disability" and the Future of Medicine'.

3
Why There is No Right to Know One's Genetic Origins

Heather Draper

Introduction

In June 2004, the Human Fertilisation and Embryology Act 1990 was modified to lift gamete donor anonymity from April 2005. This change also includes embryo donation and egg-share arrangements. The UK is not the only country to remove donor anonymity. Sweden changed its law in 1985, Austria and New Zealand in 1992, Victoria (Australia) in 1995 and The Netherlands in 2000[1]. Pressure for change seems to come from several quarters. Family law in the UK, even prior to the introduction of the Human Rights Act 1998, had been moving towards the view that children's welfare interests include knowledge of their genetic origins (Wallbank 2004). The announcement by the Department of Health Minister Melanie Johnson, reinforces this notion and suggests others:

> Donor-conceived people have a right to information about their genetic origins that is currently denied them, including the identity of their donor ... donor people should not be treated so differently from adopted people ... The interests of the child are paramount. We live in an age where ... our genetic background will become increasingly important ... information about their genetic origins ... is rightly theirs.[2]

Her statement suggests that donor people[3] ought to have the same rights to trace their genetic family as adopted people, who have been entitled to know the identity of those who placed them for adoption since 1975 (Children Act 1975, s. 26). Moreover, the need to know is linked with the potential medical benefits of having a full genetic history available. Of perhaps less significance generally, but important to affected individuals, is the occasional need for genetically related live donors of soft tissue. Having access to the donor – and even his/her wider family – increases the potential pool of suitable tissue donors. Moreover, there are those who argue that gamete donors cannot alienate themselves from parental obligations[4] though this has been disputed.[5] The Department of Health conducted a public consultation prior to the change in the law, the response to which was mixed, but many organisations and individuals (particularly donor people and organisations representing them) argued strongly in favour of a change and their views prevailed.

Resistance to the change centred on two arguments, neither of which seems compelling: first, that it would result in a decrease in the number of donors; and second, that gamete recipients are entitled to family privacy, making it wrong to require them to inform their children of their genetic origins and thereby introduce potential conflict into their family life. The first argument assumes that the utility of enabling more babies to be born by gamete donation[6] outweighs the disutility of the unhappiness caused to those donor people who would like, but are unable, to discover their genetic origins.[7] However, as one donor put it:

> [t]he fear that the removal of anonymity would lead to a drop in donors numbers is ... a secondary issue. Either something is cruel or it is not. It does not become less cruel because more babies are able to be produced by it.[8]

It is not obvious that in countries where anonymity has been lifted that donor numbers have actually fallen,[9] though the characteristics of those donating appears to have changed.[10] The issue of whether or not the number of donors will decrease is secondary to the question about whether people are harmed by being conceived by donated gametes and whether this harm – assuming that it exists – can be righted by giving them access to the identity of the donor.

Moving to the second argument against lifting anonymity, namely the right of the recipients to maintain their privacy by not disclosing to their children the circumstances of their conception. There is considerable evidence that recipients are reluctant to disclose to their children the circumstances of their conception[11] despite evidence strongly suggesting that *not* informing children can be damaging,[12] as well as evidence that their children are *not* harmed by this knowledge.[13] If children have a *right* to know their genetic origins, then there must be corresponding *duty* to tell them of the circumstances of their conception so that they can exercise their right to know.[14] Moreover, it could also be argued that the state has a duty to ensure that children are told, for instance by having this information recorded on their birth certificates. Given that recipient parents are apparently unwilling and unlikely to disclose this information, the scope for parental discretion should be removed since it impedes the rights of children to know and may adversely affect their welfare in other ways. However, three different claims need to be separated at this point. The first is that children have a right to know the identity of their donor. The second is that parents ought not to be dishonest with their children. The third is that it is better for children to know the circumstances of their conception and to have an honest relationship with their parents about these circumstances. It is possible to argue for honest relationships *and* accept that children are better off knowing the circumstances of their conception *and* still be doubtful that they have the right to identifying information about the donor.

In this chapter, I will take a critical look at the three main arguments used in support of the change in the law: that there should be parity between donor and adopted people; that there is a right to know one's genetic origins and that this right is at least in part supported by the significance of a complete genetic history for medical treatments. In so doing, a distinction will be drawn between children having a right to honesty about the circumstances of their conception and having a right to identifiable information about donors.

Gamete donation and adoption

> Donor people should not be treated so differently from adopted people ...

There are certainly some similarities between donor people and adopted people. Both may lack a close genetic connection to their parents and both may seek to know more about those to whom they are most closely genetically related. Those arguing in favour of identifiable donors depend heavily on these similarities, placing less weight on other similarities but also on the differences. Infertility and its treatment using gamete donation, like adoption, are shrouded in stigma and secrecy[15] and the 'need' to seek genetic origins is asserted as a 'natural' response that is not subject to much criticism. O'Donovan[16] argues that both the stigma and the response to stigma have a tendency to affirm the perceived norms. For instance, in the case of adoption, the stigma of illegitimacy was attached to both the birth mother and resulting child(ren). Legitimacy was the norm and preserved the prevailing moral views that attached to marriage and the place of women at the time. Likewise, stigma also attached to the adopting couple because this was a sign of their infertility.[17] The response, however, also reinforced the stigma by using the law to ensure that adopted children were able to be like 'normal' children and trace their genetic relations (or blood family) with whom they had some overwhelming or 'natural' affinity, the absence of which acted as a barrier to 'proper' relationships with their now so-called 'social' family. Assuming that O'Donovan's argument is correct, the non-stigmatising response would have been to affirm the importance of either pluralistic accounts of the family (where genetic relatedness is neither a necessary nor a sufficient condition for rights and responsibilities) or to affirm the importance of the social relationships that families represent, thereby legitimising adoption as a normal mechanism for the foundation of family life.

Arguably, the same process can be observed in the change to identifiable gamete donation, which both reinforces the stigma of infertility and affirms the view that ultimately legitimacy lies in genetic connectedness (it is only within a genetically connected family that one is not a 'cuckoo in the nest'). Giving children the right to trace their donors reinforces the view that this is the right thing to do, and reinforces the supremacy of genetic connectedness over social connectedness. Arguing that it is 'only natural' that children should want to trace their 'genetic family' reinforces the norms associated with genetic relatedness, the very norms that arguably lead to the desire to do the tracing.

Looked at in this way, the appeal to treat donor people like adopted people sounds more like an appeal that both should be treated differently from 'normal' people with a view to encouraging conformity to the norm. And both groups end up worse, rather than better, off as a result.

So much for the similarities between adoption and gamete donation – what about the differences? The most obvious one is that the child does not exist at the time of gamete donation.[18] What is given in the case of adoption is a child; what is given in the case of gamete donation is the opportunity to create a child. Of course, if we take the view that the driving force behind the desire to seek out those with whom we have the closest genetic connection is a natural or instinctive one related to the importance of genetic connectedness, then this makes little difference. But it is not obvious that this is the case when we look at adoption.

Adopted children are more likely (at least initially) to be interested in the birth 'mother' than the genetic 'father'.[19] There are many explanations for this. It may be the case that only the woman's name appears on the birth certificate. Thus, in order to stand any chance of discovering one's male genitor, one may first have to seek one's female genitor/ birth 'mother'. Alternatively, it may reflect the view that care (in an emotional as well as practical sense) of children is more the domain of mothers than of fathers, making the decision of a father to 'reject' a child – particularly an unwanted baby – more comprehensible than the apparent rejection by the mother. On this interpretation, the interest is not primarily genetic, but reflects the more social/emotional aspect of parenting. The failure of parenting resulted in the abandonment of the child, and the quest for the birth 'mother' is a quest for an understanding of this abandonment. But this is speculation and it is difficult to make generalisations about individuals' reasons for seeking out those named on a birth certificate. For instance, comparisons are sometimes made between the desire to seek the birth 'mother'/gamete donor and the preoccupation that some have with family trees or genealogies, which could also interpreted as an indication of a deep and inherent driving force to seek the genetic connections. Another interpretation, of course, is that it is simply curiosity and a hobby that interests one in one's spare time.[20] If this is so, then perhaps the 'drive' to seek the people on one's birth certificate is also mere curiosity, which is encouraged by the paper and electronic records that make the process of discovery interesting and attainable, or exacerbated by a sense that it is unjust that there is information 'out there' about oneself and to which one has been denied access.[21] O'Donovan notes that of the first 500 applications for original birth certificates, fewer than half of the applicants expressed a desire to meet or contact those named.[22] This was reinforced in a study by Howe and Feast whose participants were more interested in satisfying in their curiosity about the origins of mannerisms or physical characteristics than in forming relationships.[23]

Are there any moral differences or similarities between giving a child for adoption and giving gametes? Ought adopted/donor children expect that people who give children for adoption or donate gametes have some sense of responsibility for the children that they give up or who are created as a result? In the case of adoption, one might argue that the act of giving a child for adoption is the final act of parental responsibility for that child.[24] If this is so, and given the restrictions and safeguards in place to protect children being placed for adoption, there is no further moral responsibility, even if one retains an emotional interest. Benatar[25] has argued that donation amounts to a failure of responsibility because similar checks are not in place to ensure that the interests of donor children are protected. He argues that, unlike parents who give their children for adoption, gamete donors cannot alienate themselves from their parental rights and responsibilities because they cannot be certain that the children created will be looked after properly (he argues that the decision is made 'lightly' as a result). Bayne[26] counters that the most that this shows is that there is some procreative (as opposed to parental) responsibility for gametes, though he agrees that this responsibility can only be discharged if donors are donating within a system that does attempt to ensure that those being treated will be adequate parents.[27] However, it is not clear what this amounts to in terms of responsibilities for the resulting child. Perhaps at most donors can be expected to respond positively to an appeal for help from a donor child who claims her life is not worth living – assuming that there is something that the donor can do to improve her lot – and also to ensure that their gametes will not be used to create a child who, for genetic reasons, will have a life not worth living. This still leaves quite a large gap between the possible expectations of the child and the responsibilities of the donors (see below). It might, however, justify access to genetic information that could result in effective treatment or requests for tissue donation, where the quality of the donor person's life would be very poor without this treatment.

Both Benatar and Bayne rely, however, on the importance of genetic relatedness (via the causing to exist of a child) as the basis for these arguments.[28] If genetic relatedness is not sufficient to generate parental responsibility, then their arguments fail.[29]

Arguably, changing the law to allow donor people access to the identity of the donor creates a responsibility that would not otherwise exist, especially when such a law is not retrospective (anyone who donates from April 2005 knows that they will be identifiable and could, therefore, be deemed to have accepted this responsibility). The change in the law could itself engender in the donor person the expectation that donors *ought* to be willing at the very least to be identified, possibly more (meet with the donor person, form a relationship, introduce members of his family to her, more of which in the following section). Under this interpretation, it is the law itself that generates responsibilities rather than reflecting pre-existing responsibilities. If genetic relatedness is not sufficient to generate parental responsibilities, the law itself was unnecessary and has served to generate needless expectations.

It is counterintuitive to argue that one has greater responsibilities for a child created from one's donated gametes than for a child one gave over for adoption, counterintuitive in the sense that no child existed (and no child may indeed ever exist) at the time of donation. There is no relationship with that child such as could be forged by carrying him to term, as in the case of adoption. At their best, the genetic related-ness argument suggests that donors should demand a demonstrable and serious commitment to parenting by the *recipients* of gametes and that no child will be created whom living seriously disadvantages. Moreover, the genetic relatedness arguments tend to overlook that act of will on the part of the recipients without which the child would not exist. Granted the donor's contribution is necessary, but it is not sufficient to ensure the existence of the child. It is the recipients with whom the final responsibility for causing the child to exist rests and they undertake this responsibility as part of a personal parental project and not to aid someone else's (e.g. the donor's) parental project.

To summarise, the similarities between adoption and donation depend entirely on a view that genetic connectedness is vital – yet the analogy is drawn to demonstrate that it is vital (it has been asserted in the case of adoption, donation draws significance from its similarities with adoption). Yet there are other similarities with adoption that sug-gest that it is the stigma associated with lack of genetic connectedness and adherence to the norms of fertility (rather than genetic relatedness *per se*) that have motivated the change in law in the case of both adop-tion and donation. If this is correct, then the analogy serves to add to the discrimination of both groups rather than to promoting equality generally. The dissimilarities are also significant and would appear ironically to suggest that donor people might be better placed to argue for identifiable donors than adopted people are for identifiable birth 'parents'. My own view is that the arguments that suggest this are weak and that the change in the law has actually generated responsi-bilities for the resulting child that previously did not exist. To counter this view, it would have to be shown that genetic relatedness *per se* gives children the right of access to identifiable information; this is the subject of the next section.

Is there a right to information about one's genetic origins?

As previously stated, this supposed right seems to be supported in two ways: the importance of genetic information for medical treatment, etc. and some integral or inherent need/ interest in knowing one's genetic origins.

The importance of genetic information for medical treatment

We live in an age where ... our genetic background will become increasingly important.

Could the right to know the identity of progenitors be based on the significance of a complete genetic history for some medical treatment or the possibility of tracing willing living donors of soft tissue (e.g. bone marrow or a kidney)? This question – even assuming the significance given to genetic information that it suggests – requires some careful unpicking.

First, it is important to be clear whether the right would be conferred by virtue of some special relationship between donor and donor person based on genetic connectedness, or whether it is based on some general claim that co-operation should not be resisted when one person needs something (information/tissue) that would improve his health or life prospects that someone else (no matter who they are) can provide without undue cost to themselves. If it is a claim based on a special relationship – or obligation – the donor has by virtue of genetic connectedness, then we return to some of the problems encountered above. It must be shown that the donation of gametes entails some parental or quasi-parental obligations to the resulting individual, obligations greater than are owed to a genetic stranger. Alternatively, it would have to be shown that *anyone* who has genetically relevant information (or tissue) that would significantly improve the life or life prospects of another person should be identified to the person needing the information/tissue. For anyone attracted to the notion of a duty of rescue,[30] such a proposition would not be unpalatable. Information about all of us could be stored and our identities provided to any compatible other in need of transplantable tissue/a complete genetic history. Safeguards to the rescuer that are part the duty of rescue would, of course, apply. But we are reluctant to endorse the duty of rescue in related circumstances, for instance preferring – if our laws are the measure – to accept that people will die for the want of an available organ before endorsing policies that would require the 'donation' of all useful cadaveric tissue. It may even be argued that the right to privacy trumps the right to know in such cases. Likewise, there has been a resistance to the view that genetic information can be shared with other family members without consent, for instance, where one family member is known to have a condition that is likely to affect other family members and which is either treatable or could be ameliorated if treated soon enough following regular screening. This reluctance does not seem compatible with a justification for identifiable donors based on the increasing importance of genetic information/history.

Again, and in my view more importantly, knowing the *identity* of the donor is a not prerequisite for having one's medical needs met. The relevant history could be taken and made available, or the request for tissue could be made, without the recipient knowing the identity of the donor by using intermediaries. In the case of tissue donation, the use of anonymous donors might be unusual in living tissue transplantation, but it is the norm in cadaveric donation of tissue. The claim of the donor person to know genetic information for medical reasons must be considered against the background of existing arguments for and against requiring family members to share genetic information. The general argument against requiring family members to disclose genetic information rests on the claim that it is an intrusion of privacy: the known-to-be-affected family member would be identifiable to the likely-to-benefit family member. In the case of donors, a family history – even one that included the wider family – could be taken without intruding on privacy if 'invasion of privacy' is taken to mean 'information can be traced back to someone who can be identified' *because* they are not members of the same social family.[31] Given the problems that arise in sharing genetic information in conventional families, the identification of donors could actually be counterproductive in the quest for a genetic history, as the anonymity may prove attractive to the donor and his family as a mechanism for protecting genetic privacy. Either way, the case for the donor person having the *right* to identifiable information (where 'right' suggests access with or without consent) is only as strong as the case for *anyone* having such information, whether donor-conceived or not.

Thus, whilst a case can be made that donors should be identifiable to *someone* – i.e. traceable by an intermediary – on the strength of the possible future needs of donor people, it is not clear that donor people can claim to have a right to identifying information on the basis that they may need genetic information or tissue that is vital to their health or life prospects. It might be more sensible to take as full a genetic history as possible, and a blood sample for future use, at the time of donation. This would also help in cases where the donor is dead or cannot be traced at the time of need.

The inherent importance of knowing one's genetic origins

[I]nformation about their genetic origins ... is rightly theirs.

The importance of knowing one's genetic identity is often assumed or asserted without justification as if there is an innate drive beyond the control of the will to know.[32] When pressed, people making the assertion will point to inherited characteristics (like being quick-tempered, disorganised or generous) or physical characteristics (facial shape, hand size or other general physical similarities) and note that it is good to know where these come from (in the case of characteristics) or good to have the sense of belonging that is generated through shared physical similarities. It is not clear why it is important, for instance, to know that one is quick-tempered like one's father and his father before him, unless this suggests that it is something that one needs to guard against. The question of the impact of genetic knowledge on freewill is too large for this chapter, but we seem to be a long way from supposing that one could argue that one has no control over one's temper if one can show that being quick-tempered is something one inherited from one's father. Even if this were to become a legal defence, the argument that donor people had a right to know their genetic origins for this reason would not show that it is inherently important for them to know, but rather that they had the same right to a fair trial as anyone else.

Nor is it obvious that our fundamental or most important senses of belonging come from shared physical traits: religion, for instance, is not inheritable but tends to be shared within families, and religious ceremonies cement important family and community relationships that lead to feelings of belonging. Moreover, we should not discount the effects of nurture. Compassion, commitment to political, moral or cultural ideologies, abusive relationships and decency, etc. can also run in families and cement a sense of belonging and identity. It is not clear why genetically defined traits are more important or more decisive than such socially acquired traits. The importance of genetic relatedness seems to come from the fact that – even before decisive DNA testing – it was in some sense provable. For instance, some men want to know that children are the product of their loins before taking paternal responsibilities for them. From an evolutionary perspective, it could be argued that it is important that individuals only invest in children that perpetuate their own – or very similar (e.g. their siblings') – genes. But evolution is not just about individuals it is also about communities.[33] For communities to survive, and therefore for the individuals within them to have a good chance of passing on their genes, it is also necessary to have individuals who are prepared to invest in non-related children, so that the community as a whole can survive the ravages of war, famine or other social and family upheaval. Again, it is beyond the scope of this chapter to summarise further the individual versus social notions of evolutionary survival; the point is that it is not obvious that the drive to ensure or determine genetic connectedness is either innate (necessary) to human survival, nor that it is more than a convenient (because provable) method of ensuring justice for men.

As we have seen, part of the answer to the question about the importance of genetic connectedness seems to lie in the obligations that it might be supposed that we have to genetically related children. Few people would argue that genetic relatedness is necessary to generate parental obligations – it is accepted that such obligations can be generated legally (e.g. through adoption) and perhaps even morally (through creating an expectation in children that they will be forthcoming, perhaps through social practices like encouraging the title 'mummy' or 'daddy' outside formal adoption). More controversy is generated by the suggestion that a genetic connection is sufficient for parental obligations to follow (e.g. discussion of Benatar and Bayne above) whether or not the concept of parent is essentially pluralistic. By and large, we accommodate a pluralistic notion of parenting and parental obligation that includes social connectedness as a vital element (rare exceptions aside, one is not a parent in any meaningful sense if one willingly has no social contact with one's children whether or not they are genetically related). The significance of genetic connectedness is most prominent when there is some dispute: when someone is trying to assert or deny parenthood[34] (e.g. when a man wants access to a child or wants to deny responsibility for a child) or when parenthood is asserted by more than one party or set of parties (e.g. deciding who the parents are when a surrogate wants to keep or regain a baby). The importance of the social aspect of parent-child relationships is ironically highlighted when donor people assert the right to identifying information, as opposed to the right to honest dealing with their parents (the right to know the circumstances of their conception). In many cases, what is sought is not the simple knowledge that one is not – as one might reasonably suppose – genetically related to one's parents, but the right to be able to identify precisely those with whom one's closest genetic connections lie with a view to forming some social connection, even relationship. Some donor people, for instance, want to meet the donor at least once and perhaps also donor grandparents or siblings too.[35] Thus, it is not the *identity* of the donor that is at stake here, but rather a possible relationship with the donor and his/her existing family. This need seems to be supported by European Convention Article 8. This grants the right to respect for privacy and *family life* (my emphasis). The phrase 'family life', although flexible, is a richer notion than the possession of a name or even a face to a name, and it is this Article that forms at least part of the basis for the claim that there is a legal right to identifiable information.

Is it reasonable for donor people to assert, then, not the right not to be deceived about their genetic origins but the right to have a family life with the donor? Part of the answer to this question lies, I suspect, in the question of how many families we are entitled to have and the circumstances under which genetic family ties are disrupted.

If a child is forcibly removed from good parents for a political reason – compulsory removal of fair-skinned Aboriginal children in Australia in the nineteenth and twentieth centuries, for instance – then it is not unreasonable for those parents to claim a right to the return of that child, or for the child to be considered to have the right to live with those parents. The removal is unjust by any definition. But it is not unjust because of genetic connectedness; it is unjust because the parents did nothing to deserve the removal, upheaval in life, grief at the loss of loved ones and a loss of family life. It would be equally unjust if the child concerned had been formally or informally adopted and was similarly removed for no good reason. It is not clear that adoption *per se* or donation amounts to the same kind of injustice. No one under UK law or policy has their children or gametes forcibly removed (unless in the case of the former they have proved themselves to be unfit parents – which arguably is not unjust).[36] It is true that very young children can be given up for adoption without their consent; thus such children could later argue that from their point of view their family, etc. was disrupted unjustly, but it is not clear that babies adopted at birth are traumatised by the experience or fail to bond with adopted parents, nor are such babies autonomous beings, so the basis for the claim of being wronged by the adoption is not clear. However, in the case of gamete donation, the child-to-be does not exist at the time of donation and, therefore, has no bond with the donor that is disrupted by traumatic separation. What donor people would have to show, then, is that the process of donation is seriously unjust to them in later life, without simply basing this argument on an assumption that genetic connectedness is sufficient grounds for asserting this. If this is so, then we return to the position outlined by the sperm donor above: if it is harmful, why do we permit the use of donor gametes at all? This is an argument for not allowing donation, not an argument for identifiable donation. And if it is the case that having a later social connection with the donor ameliorates this harm, then the law should be retrospective – covering all donor people, not just future ones – and go much further than granting access to identifiable information. It should require donors to form family relationships with donor people.

Before looking at how reasonable this suggestion is, a further point needs to be made that relates to the distinction between donor people and donor children. Clearly, minors require special protection and it is young children that are most immediately created by the process of donation. If young children are harmed by not being told the circumstances of their conception, as the research referred to above suggests, then a case – beyond the obvious principled one that it is better to be honest than dishonest – can be made for requiring that adoptive and recipient parents are honest with their children. But it is as adults that the right to identifying information for donor people is bestowed; there is no such legal entitlement before the age of majority. Yet as adults donor people do not require the special protection afforded to children, and referring to donor people as donor children and in the same breath talking about the right of the child, confuses the issue.[37] If children are – all things being equal – best brought up by genetically related parents, then they should be 'returned' to the donors as children, babies even, which rather defeats the objective of infertility treatment and again suggests that the practice of gamete donation should be discouraged. As adults, the needs and interests of donor people, the donors and the existing parents are more equal, and it is difficult to see why donor people's needs should be permitted to trump those of everyone else concerned, and further justification would be needed for these needs to trump the interests of any minors likely to be affected (for instance, those in the donor's existing family). In this context, the assumption of the utilitarian calculus explored above gains further weight: would donor people rather not exist than exist as donor people? If there is no injustice in bringing into existence donor people, when the alternative is for them not to exist at all, then it is difficult to see what the overwhelming harm is that the change in the law seeks to right. If we agree that there is a wrong, then it should not continue to be repeated and use of donor gametes should cease forthwith and the law should be retrospective.

We can now return to the issue of whether donors ought to be willing to enter into a family relationship with donor people, or put another way, whether donor people are entitled to a family relationship with donors. The distinction here is between genetically related people *voluntarily* entering into a relationship – perhaps like those entered into by those who trace their wider genetic family though genealogy and contact them, or like the relationships that are sometimes rekindled through Friends Reunited – where both parties are willing to meet, and, on the other hand, the sense that a relationship is *owed* or required by virtue of the donor being the genitor of the donor person. We have already noted that donor people are not at this point dependent children whose needs should be met because of their vulnerabilities, but nevertheless, parents do feel obligations for their grown-up children, so different kinds of obligations could be forthcoming that might normally be expected in families. The question that remains is why it is the *donor* who owes these obligations and not the donor person's existing parents. Is the donor person entitled to expect such obligations, or even love, affection and so forth from the donor when she already has a family life?[38] Do people have the right to family life with more than one family? The answer here seems to be that it would be wrong to prevent someone having more than one family – or extending their family – but it is not obvious that they are *entitled* to extend their family to the extent of recruiting unwilling members. Nor is it clear that one person can be required to love another. On this argument, there is no reason to prevent willing donors and donor persons from having such a relationship anymore than it would be reasonable to prevent non-genetically related persons entering into quasi-parent/child relationships as adults. But in this case, connecting donors and donor people would be more like Friends Reunited where parties willing to re-establish contact make these wishes known and if those they seek are also willing they look in the same place for their contact details, but this does not give either side the right to the identity and contact details of the other, let alone the right to insist upon a relationship.

A right for a few or a right for all?

I have argued that because a genetic connection is not obviously sufficient to establish parental responsibilities or the right to family life, donor people do not have the right to identifying information about donors, but I do not expect everyone to find my view decisive. If so, then what are the other consequences of adhering to the view that a genetic connection is sufficient? One such consequence has already been established, namely that it is unjust that the UK law – and the laws elsewhere – are not retrospective, preventing those who were conceived prior to the change in the law, or from gametes, etc. donated prior to the change in the law, from having the same rights. Another is the question of whether it is correct to continue to create children in this way.

But there is an additional consequence, namely that this is a right that should be given to *all* people: being in a position definitively to establish one's genetic origins should be granted to *all* adults. This suggests that it is not unreasonable for children to insist on testing their parents and, if these tests show that they are not genetically related, to be entitled to the identity of those to whom they are related and to test such persons to establish the truth. Moreover, for those who argue that one cannot be alienated from one's rights (as enshrined in conventions like the UN Convention on the Rights of the Child or the European Convention on Human Rights, or the UN Declaration on Human Rights) such tests and disclosures should be mandatory. To argue that they ought not to be mandatory concedes, or so it seems to me, the argument about sufficiency. If one is free to determine for oneself who one's family is, then this suggests that genetic connectedness is not sufficient to determine family relationships and gives at least equal weight to social rather than genetic connectedness.

Notes

1 See L. Frith (2001) 'Ethical and Legal Debate', *Human Reproduction* 16(5), pp. 818–24.
2 Available online at: http://www.dh.gov.uk/PublicationsAndStatistics/ PressReleases/PressReleasesNotices/fs/en?CONTENT_ID=4070524&chk=jLTII %2B.
3 I dislike the term 'donor people', just as I dislike the term 'adopted people'. It should be clear from this chapter that I disapprove of the view that people are, or should be, defined by the processes of their conception or circumstances of their birth. However, 'people conceived by gamete donation' and 'people who were adopted as a child' are long and cumbersome phrases and it seems equally distasteful to refer to the groups concerned as PCGD and PAC respectively.
4 For example, D. Benatar (1999) 'The Unbearable Lightness of Bringing into Being', *Journal of Applied Philosophy* 16, pp. 173–80; J. L. Nelson (1991) 'Parental Responsibilities and the Ethics of Surrogacy: a Causal Perspective', *Public Affairs Quarterly*, 5, pp. 49–61.

5 For example, T. Bayne (2003) 'Gamete Donation and Parental Responsibility', *Journal of Applied Philosophy*, 20, pp. 77–87; G. Fuscaldo (forthcoming) 'Genetic Ties: Are They Morally Binding?', *Bioethics*; H. Draper (submitted) 'Fatherhood, Responsibility and Genetics: Disentangling Genetic Relatedness and Paternal Responsibility', *Journal of Medical Ethics*.

6 Assuming that this resolves the misery of infertility previous experienced by their parents, and assuming that for the most part these children will have relatively happy lives.

7 For few would argue that the lives they have, even if marred by this lack of knowledge, are not on balance worth living for this reason.

8 A. McWinnie (2001) 'Gamete Donation and Anonymity', Human Reproduction 16(5), pp. 807–17.

9 There is some controversy over the extent to which lifting anonymity reduces the number of donors. Gillian Lockwood, Director of the Midland Clinic, for instance argued at a Progress meeting in London (17 November 2004) that the falling numbers of donors in European countries that had already moved to identifiable donors was masked by the fact that couples were now seeking treatment in other European countries that still maintained anonymity. By October 2004, however, BBC News had already begun reporting on a sperm donor crisis. See http://news.bbc.co.uk/1/hi/wales/south_east/3753850.stm.

10 Lalos et al. (2003) 'Recruitment and Motivation of Semen Providers in Sweden', *Human Reproduction* 18: 212–16; E. Blyth (2002) 'Information on Genetic Origins in Donor-assisted Conception: Is Knowing Who You Are a Human Rights Issue?' *Human Fertility*, 5, pp. 185–92.

11 J. Robinson et al. (1991) 'Attitudes of Donors and Recipients to Gamete Donation', *Human Reproduction*, 6, pp. 307–9; C. Gottlieb et al. (2000) 'Disclosure of DI to the Child', *Human Reproduction*, 15, pp. 2052–6; S. Klock et al. (1996) 'A Comparison of Single and Married Recipients of DI', *Human Reproduction*, 11, pp. 2554–7; R. Cook, S. Golombok, A. Bish and C. Murray (1995) 'Keeping Secrets: A Controlled Study of Parental Attitudes towards Telling about Donor Insemination', *American Journal of Orthopsychiatry*, 65, pp. 549–59; A. Brewaeys et al. (1993) 'Children from Anonymous Donors: An Inquiry into Heterosexual and Homosexual Parents' Attitudes', *Journal of Psychosomatic. Obstetrics. & Gynaecology*, 14, pp. 23–35; A. Brewaeys et al. (1997) 'DI: Child Development and Family Functioning in Lesbian Mother Families', *Human Reproduction*, 12, pp. 1349–59; V. Soderstorm-Antilla et al. (1998) 'Embryo Donation: Outcome and Attitudes among Embryo Donors and Recipients', *Human Reproduction*, 16, pp. 1120–8.

12 A. Clamar (1989) 'Psychological Implications of the Anonymous Pregnancy', in J. Offerman–Zuckerberg (ed.), *Families in Transition: A New Frontier* (New York: Plenum Press), pp. 111–20; R. Snowden, (1993) 'Sharing Information about DI in the UK', *Politics and the Life Sciences*, 12, pp. 194–5; K. Daniels and K. Taylor (1993) 'Secrecy and Openness in Donor Insemination', *Politics and the Life Sciences*, 12, pp. 155–70; Donor Conception Support Group of Australia Inc. (1997) *Let the Offspring Speak: Discussions on Donor Conception* (Georges Hall, NSW: Donor Conception Support Group).

13 C. Turner (1993) 'A Call for Openness in Donor Insemination', *Politics and the Life Sciences*, 12, pp. 197–9; S. Golombok (1998) 'New Families, Old

Values: Considerations Regarding the Welfare of the Child', *Human Reproduction*, 13, pp. 2342–7; K. Vanfraussen, I. Ponjaert–Kristoffersen, and A. Brewaeys (2001) 'An Attempt to Reconstruct Children's Donor Concept: a Comparison between Children's and Lesbian Parents' Attitudes towards Donor Anonymity', *Human Reproduction*, 16, pp. 2019–25.

14 L. Frith (2001) 'Beneath the Rhetoric: the Role of Rights in the Practice of non-Anonymous Gamete Donation', *Bioethics*, 15(5), pp. 473–84.

15 K. O'Donovan (1991) '"What shall we tell the children?" Reflections on Children's Perspectives and the Reproductive Revolution', in R. L. Lee and D. Morgan (eds.), *Birthrights: Law and Ethics at the Beginnings of Life* (London: Routledge), pp. 96–114.

16 O'Donovan, '"What shall we tell the children?"'.

17 Even though not all of those adopting children are actually infertile.

18 Unless the donation takes the form of an embryo.

19 Again, the terms 'birth mother' 'genetic father' and even 'mother' or 'father' are used for convenience here. For instance, if the 'genetic' father wanted nothing to do with the child, then there is a real sense in which he is no kind of father at all, and is actually unworthy of a title that suggests the fulfilment of certain responsibilities to children (as well as rights) that he has failed to deliver.

20 After all, books and magazines on the topic or more likely to be found in the hobbies section of the store than in the health or psychology sections.

21 Again, a distinction can be drawn between the right to know one's genetic origins and the right to know information about oneself that is known to others but not to oneself. In the case of the latter, one might be reasonably annoyed that one is denied access to the information on the basis of transparency or freedom of information as distinct to its being related to one's genetic origins.

22 Only 46 gave locating a genetic parent or relative as the primary reason. O'Donovan, citing C. Day (1979) 'Access to Birth Records', *Adoption and Fostering*, 3(4), p. 17.

23 D. Thowe and J. Feast (2000) *Adoption, Search and Reunion: the Long-term Experience of Adopted Adults* (London: The Children's Society).

24 To be distinguished from having one's children removed from one's care and placed for adoption, which might be regarded as a failure of responsibility so grave that responsibility is removed by the state, along with the child.

25 Benatar, 'The Unbearable Lightness of Bringing into Being'.

26 Bayne, 'Gamete Donation'.

27 And he considers that the current system of regulation in the UK is such a system, even though it is not as rigorous as the UK adoption system.

28 Though for Bayne, this only amounts to part of a more pluralistic account of paternity/maternity. On this point cf. David Oderberg, chapter 5 in this volume.

29 Draper, 'Fatherhood, Responsibility and Genetics'.

30 For an introduction to duty of rescue, see T. Beauchamp and J. Childress (1994) *Principles of Biomedical Ethics* (Oxford: Oxford University Press), pp. 266–8.

31 We should be arguably less concerned about information *per se* about us being in the public domain when the information about us is stripped of anything that could lead us to be identified. The distinction between anonymous and identifiable participants is, for instance, is used to support epidemiological research and is an interpretation that seems to be supported by the Data Protection Act 1998. See also W. Rogers and H. Draper (2003) 'The Ethics of Confidentiality in Medical Ethics', *Journal of Medical Ethics*, 29, pp. 220–4; and H. Draper and W. Rogers (2005) 'Re-evaluating Confidentiality: Using Patient Information in Teaching and Publications', *Advances in Psychiatric Treatment* 11: 115–24.

32 The UN Convention on the Rights of the Child, Article 7 assumes the fundamental importance of the right of the child to know its parents. As Frith ('Beneath the Rhetoric') notes, the Convention was not formulated with donor children in mind. Perhaps it had in mind cases where children are forcibly removed from their parents for no other reason than that someone else is deemed more deserving of children. Note too that the Article, and its application through the European Convention on Human Rights, Article 8, assumes that 'parent' means 'genetically connected person' – that genetic connectedness is sufficient for parental rights and responsibilities, something that requires justification.

33 My thanks to Moli Paul for bringing this point to my attention.

34 I use the term 'parenthood' rather than 'paternity' because this could apply equally to women.

35 McWinnie, 'Gamete Donation and Anonymity'.

36 I concede that in the past unmarried women were subjected to huge pressure to give up their children for adoption, and that such coercion would make them a special case.

37 One illustration of this confusion is found in the concluding comments of McWinnie ('Gamete Donation and Anonymity', p. 816): 'We have accepted the United Nations Declaration of Human Rights and signed up to the UN Convention on the Rights of the Child (1989) with its emphasis on giving children a voice and that their voice should be heard. Donor offspring do not consider that their voice has been heard, or their perspective given any consideration, or that they have been fairly treated.' This statement gives the impression that it is the voice of children in the sense of *minors* that has not been heard rather than children in the sense of offspring who are not actually minors at the time of speaking.

38 For the sake of argument, let us suppose that she is not an orphan, though if she were, it is not clear why it should be the donor who is charged with compensating for this tragedy.

4
Compromise and Moral Complicity in the Embryonic Stem Cell Debate[1]

Katrien Devolder and John Harris

In September 2004, Italy's health minister, Girolamo Sirchia, hailed the successful treatment of a five-year-old boy with thalassaemia, an inherited form of life-threatening anaemia. The therapy involved transplanting stem cells of the umbilical cord blood of the boy's newborn twin siblings. The minister hoped to use this case to convince the Italian public of the potential of non-embryo-derived stem cells and to justify the contentious Italian law on assisted reproduction. However, soon after his 'triumph' it became known that the twin pregnancy was realised with IVF and the selection of embryos through PGD and HLA typing, in a hospital in Turkey, techniques which Sirchia considers as immoral and which are outlawed by the Italian government.[2]

Promises of stem cell research

Stem cells hold great promise for medicine. Because of their proliferation and differentiation capacities they are widely believed to represent the greatest promise for medicine in the twenty-first century. They could be used to treat a variety of diseases and conditions, including diabetes, Parkinson's disease, heart disease, spinal cord injuries, blindness, deafness and many types of cancer. Unlike current drugs, which mainly delay symptoms of diseases, stem cells may make it possible to replace damaged tissue or even whole organs. Their value extends beyond utility in cell therapy. Stem cells are also useful tools for fundamental research, mainly for gaining knowledge about early human development, cell division and differentiation mechanisms, gene and protein function, for developing drugs and toxicity testing, and for developing models of human diseases.[3] This immense promise gives powerful moral reasons for pursuing stem cell research.

At present there are three main lines of stem cell research:[4] on stem cells originating from early *in vitro* embryos made available as surplus to those required for infertility treatments or especially created for research through IVF or nuclear transfer (embryonic stem cells); on cord blood-derived stem cells; and on stem cells originating from tissues or organs from foetuses or organisms after birth (adult stem cells). There is a growing consensus among scientists worldwide that all these lines of research are promising. Embryonic, cord blood and adult stem cells may have different qualities and might be useful for different purposes. In some cases the best option may be combined adult and embryonic stem cell therapy.[5] So rather than opting for one line of research, the ideal research strategy to reach the intended research goals would be to proceed with research on all types of stem cells.

However, embryonic stem cell research which involves the destruction of the embryo is opposed by those who regard the embryo as in some important sense 'one of us'. In the opinion of those who take this view, the embryo should never be used as a mere means, even if this could save millions of lives.[6] They advocate the legal prohibition of stem cell research or therapy that involves 'the killing' of human embryos. This approach may be summed up in the phrase 'the embryo is one of us', which entered the vocabulary of the moral status of the embryo via a leading Italian Catholic commentator. It has since become associated with this debate not only in the Italian context but more generally.[7]

Stem cell policy

The fact that there are no cures as yet on the basis of embryonic stem cells (ES cells) can appear as an embarrassment for those seeking clear justification of ES cell research. This creates a situation in which tensions between ethics and politics are unavoidable and very visible. An argument that always surfaces in the stem cell debate, and about which there is a broad consensus, is that there is no need to permit more ethically contentious ways of generating stem cell lines, if the same benefits can be realised using less contentious stem cells.[8] It is sometimes referred to as the principle of subsidiarity[9] and it is closely bound up with the principles of proportionality and necessity, which require that any action should not go beyond what is necessary to achieve the objectives. The principle is referred to very often in reports and recommendations on stem cell research. The Belgian law on the Protection of the Embryo in Vitro, for example, states in Art.3 §6 that research on embryos is allowed if no other research method exists that is equally effective and in Art.4 §1 that the creation of embryos for research is forbidden except when the goal of the research project cannot be reached through research on supernumerary embryos. The U.S. National Bioethics Advisory Commission wrote in the stem cell report of 1999 that 'the derivation of stem cells from embryos remaining following infertility treatments is justifiable only if no less morally problematic alternatives are available for advancing the research'.[10] The German Stem Cell Act stipulates that ES cell research shall not be conducted unless the scientific knowledge to be obtained cannot be expected to be gained by using cells other than ES cells.[11]

Those who use the principle that we have to opt for the least offensive moral approach assume that there exists a sort of hierarchy of the various methods of obtaining stem cells according to their ethical acceptability, on which the majority of people agree. Adult stem cells are placed at the top and stem cells harvested from embryos created solely for research purposes at the bottom. In between there is a variety of intermediate positions. If the intended goals of stem cell research can be reached with stem cells from all the possible stem cell sources, then – following this argument – only the less controversial source should be used. The question as to whether this argument is valuable and useful in a bioethical context we leave out of consideration in this chapter.[12] More important in this context is the fact that it actually serves as a guiding principle in political decisions about stem cell research. Consequently, if there was scientific certainty that adult stem cells were the most advantageous stem cells or as advantageous as ES cells to reach the intended goals, it is very likely that in many jurisdictions, policy-makers would legalise their use and would prohibit research with ES cells.

But, as already noticed, there is still scientific uncertainty about which line of research is the most promising for specific purposes. Many scientific publications have presented findings about the great potential of ES cells[13] and a majority of scientists claim that the best way to make scientific progress is to do research on all types of stem cells.[14] This presents many countries with difficult decisions about the regulation of ES cell research.

In a democracy it is always necessary to balance a wide range of interests and opinions. Most countries don't want to block ES cell research because of the potential health benefits and the contribution to scientific progress that stem cell research is believed to represent. For countries with restrictive laws with regard to the protection of embryos in the context of abortion, IVF and foetal tissue research, the problem is to justify certain forms of ES cell research without violating the spirit of these existing laws. These difficulties explain why the focus of the current stem cell debate is on the question whether ES cell research should be allowed – or federally funded[15] – or not, and, if so, which aspects and with which constraints. It is partly a re-ignition of the older debates, but there are differences that make the debate more complex, namely that the ES cells can be cultured and proliferated in the laboratory on the one hand, and that spare embryos and embryos solely created for research can be used as a stem cell source on the other.

Moral positions in the ES cell debate

Two polar moral positions can be distinguished in the ES cell debate. First, those who radically oppose research with stem cells obtained from human embryos. In their opinion, the embryo has the same moral status as a person and, consequently, has to be (legally) protected as a person. They portray adult stem cell research and other research strategies as viable alternatives.[16]

Second, there is the viewpoint that young *in vitro* embryos, regardless of their origins, may be used for scientific research on the condition that the embryos used in the experimentations will not be implanted in the womb afterwards (unless the research has therapeutic value for the embryo and the person that will result from it). The principal requirements regarding ES cell research are related to safety and control, proportionality, commercialisation and informed consent issues. In Europe, Belgium, the UK and Sweden (and to a certain extent the Netherlands) have regulations compatible with this viewpoint.[17]

Most countries adopt an 'intermediate view'. They search for a 'happy medium' between these two polar views. This middle position has two main variants: one based on the 'use-derivation distinction' and one on the 'discarded-created distinction'.[18] The first makes a moral distinction between the use of ES cells and their derivation from in vitro embryos; the second between the use of spare IVF embryos for research and the use of embryos solely created for research. Within these two positions there can again be variations. These moral positions correspond with varying legal approaches, from the prohibition of ES cell research to the legalisation of the creation of human embryos solely for research purposes.[19]

Countries that prohibit ES cell research

In Italy, the law on assisted reproduction prohibits ES cell research[20] and the Vatican has a powerful influence on Italy's policy-making. The Catholic Church has a major impact on Ireland's legislation as well. Ireland is the only European country that defends 'the right to life of the unborn' in its Constitution.[21] ES cell research is strictly forbidden, as is the case in Austria[22] and Norway.[23] Costa Rica, Brazil and Ecuador ban ES cell research and the United States currently ban federal funding for such research. Only a minority of countries worldwide strictly prohibit research on human ES cell cells.[24]

The question arises as to what these countries will do if life-saving therapies based on ES cell research were to be developed in less restrictive and permissive countries. There are two major options. A first one is that the government would not change its policy. Citizens would have to travel to countries or jurisdictions where life-saving ES cell therapies are allowed. This phenomenon is already known in the context of assisted reproduction and has been called 'reproductive tourism'.[25] Similarly, the strict prohibition of ES cell research could lead to 'stem cell tourism'. This option may not be very plausible since the chance is very high that the pressure of the public to have access to life-saving treatments that are available in other countries will be so great that governments will have to change their policies. This is already happening in some countries with restrictive regulatory environments. In Italy, the health minister has been put under pressure to change the strict legislation on assisted reproduction, which makes it harder for infertile people to get treatment and which endangers the life of women in order to protect the embryos. More than one million Italians have signed a petition to relax the law. If the court approves the list, the government will have to accept the referendum.[26]

After the scandal with health minister Sirchia who had hailed a successful non-embryonic stem cell treatment of a boy with thalassaemia, but 'forgot' to mention that the treatment was based on techniques outlawed by his policy, politicians and scientists have demanded his resignation. In Ireland similar lobbying to relax the strict laws on embryo research has occurred. Mary Harney, whose role is similar to that of a vice president, has voted in favour of an EU framework for research on ES cells and she gets support from a great many members of the scientific community in Ireland who believe that the wider therapeutic benefit of ES cell research must be considered and who have called for a wider informed debate on the merits of ES cell research.[27] (However, in the case of 'abortion tourism' no changes thus far have occurred.)

A second policy option would be to adopt one of the compromise positions some other countries have based their policies on, for example, by allowing ES cell research to a certain extent and under certain conditions in order to reach the intended research goals. This would be in accordance with the principle that we should always opt for the least controversial approach to reach the objectives.

Two questions arise. First, can this be done without being morally complicit with the evil deeds – the killing of embryos – committed in other countries or jurisdictions? The second question is of a normative nature: once one allows some ES cell research on the basis of health and research benefits – does one have compelling reasons not to allow more aspects of ES cell research for the same purposes? We will start with the first question.

Moral complicity of countries that adopt the use-derivation distinction

Most countries don't want to block ES cell research and want some research to proceed, but under strict conditions. Some of these countries justify ES cell research on the basis of the 'use–derivation distinction'. Proponents of the use–derivation distinction make a moral distinction between the use of ES cells for research on the one hand and the derivation of these cells from the human embryo on the other. Using ES cells for research is considered ethically acceptable, while their derivation is not, since it involves the killing of 'one of us'.

Knowing that the defenders of this viewpoint think the *derivation* of ES cells is immoral, we can ask on what grounds they might consider the *use* of the cellular products remaining from such acts as ethically acceptable.

A first argument – or rather a necessary assumption for those who consider embryos as 'one of us' – to justify the *use* of ES cells is that they are not equivalent to embryos, so that research with those cells, as opposed to their derivation, is not problematic because it is not considered as human embryo research.[28] The underlying idea is that ES cells are *pluripotent*, that is, they have the potential to generate any cell type in the body but not a whole organism, whereas embryos are *totipotent*, which means they can develop into a new organism. A human embryo is generally defined as an entity with the capacity to develop into a human being. In a legal decision, the general counsel of the US Department of Health and Human Services stated that: 'the statutory prohibition on the use of funds appropriated to HHS for human embryo research would not apply to research utilising human pluripotent stem cells because such cells are not a human embryo within the statutory definition'. She concluded that, consequently, federal support could be given for research that uses stem cells derived from embryos by private funds. [29]

Stating that ES cell research is not equivalent to embryo research is, of course, not sufficient to justify the use-derivation distinction. One has to argue why it is ethically acceptable to use products that were obtained through an allegedly wrongful act, involving the derivation of ES cells and the killing of those who are 'one of us'. In doing so, one cannot avoid the issue of moral complicity.

Moral complicity

The central question in the complicity argument is whether benefiting from another's wrongdoing effectively makes one a moral accomplice to their evil deeds. With regard to moral complicity at the state level in the context of ES cell research this question runs as follows: to what extent is a government which legalises research with ES cells responsible for or complicit in the death of the embryos from which these cells are derived? According to their answer, individuals strongly opposed to embryo research might come to diverse conclusions about the acceptability of using ES cells. Whether one considers somebody as morally complicit in the killing of embryos depends on his/her view on the separation principle, which, applied to ES cell research, states that the act through which the cell products are obtained should be completely separated from the use that is made of these products. [30] If so, then those who use the cells cannot be considered to be moral accomplices in the act of derivation. This principle has been one of the central issues in debates on foetal tissue research[31] and in the use of vaccines developed from foetal material.[32] According to some people, the use of ES cells can be separated from the act of isolating the cells and the subsequent death of the embryo. But others say there is no real separation. We will consider both viewpoints.

Those who think the separation holds

The American Association for the Advancement of Science and Institute for Civil Society state in their stem cell report that 'many individuals are convinced that not all acts benefiting from other's wrongdoing are morally impermissible, so long as one is not involved in the wrongdoing and one's own acts do not foster, encourage or lend support to it'.[33] The relevant question to ask is whether the use of ES cells fosters, encourages or lends support to the killing of embryos.

Those who think the separation holds respond that those who *use* ES cells do not encourage those who destroy embryos unless they expressly authorise the creation of an embryo for this purpose.[34] This is why countries that allow the use of ES stem cells only approve of the use of ES cells taken from left over embryos that were originally produced for the purpose of satisfying a wish for a child by IVF.[35] Ronald Green argues that there need to be no causal connection between the use of ES cells and their derivation because at the present time and in foreseeable future, embryo destruction is entirely independent of ES cell research and therapy because surplus embryos are routinely created in the practice of infertility medicine. Therefore, the argument goes, the use of ES cells is morally closer to the use of a pancreas donated from the parents of a teen murdered in gang violence than it is to anything approximating encouragement to murder.[36]

A concern exists that some parents and clinicians may justify the creation and destruction of embryos by virtue of their beneficial uses or, more generally, that this kind of research would lead to a diminishment of societal respect towards early human life.[37] However, defenders of the use-derivation distinction deny these concerns for several reasons. First, they say, there is little evidence that the use of embryos for scientific research and the benefits this brings for medicine are a motive for parents opting not to use their embryos in a parental project. In most countries the prospect of (privately funded) research has always been present and so doesn't have new influence on the way people decide about the destination of their surplus embryos.[38] Second, it is said that the demand will not be so great as to have significant quantitative effects internationally (in case the ES cells are obtained through import, as in Germany)[39] or in the private sector (in case they are obtained from the private sector, as in the US where a clear boundary is drawn between private and public research) owing these other companies' own research interest.

Some, however, are not convinced by these arguments and think the separation cannot be guaranteed because one cannot be sure that the benefits of ES cell research will induce some couples to create gametes for research, or to give their spare IVF embryos to researchers, rather than to include them in a parental project.[40] President Bush tried to overcome these concerns by his policy decision of 9 August 2001 to allow federal funds to be used for research on stem cells from spare IVF embryos under certain conditions (Bush's eligibility criteria) and – most importantly – on the condition that prior to his announcement the derivation process had already been initiated, in Bush's words 'where the life and death decision has already been made'.[41] Federally funded researchers in the US can only make use of the 66 stem cell lines listed in the NIH's Human Embryonic Stem Cell Registry. Bush explained that 'this allows us to explore the promise and potential of stem cell research without crossing a fundamental moral line, by providing taxpayer funding that would sanction or encourage further destruction of human embryos that have at least the potential for life'.[42]

Germany considered the US's compromise position, which intended simultaneously to protect the rights of the unborn and the freedom of research, as a valid solution to solve the heated discussions preceding the policy decisions on stem cell research in Germany. The Embryo Protection Act of 1990 forbids embryo research that is not for the benefit of the embryo itself, but the Stem Cell Act of 28 June 2002 allows the import of ES cells under similar conditions as set up in Bush's decision, most important in this context, that the ES cells were derived before the 1 January 2002. The underlying idea is that 'the stem cells were produced abroad … and the "consumption" of the embryos is already finished and irreversible before they are imported'.[43]

The main argument underlying these 'compromise regulations' is that 'thanks to this policy' Americans and Germans – potential patients as well as other stakeholders – are permitted to enjoy the benefits of previous wrongful deeds, without encouraging in any way the repetition of similar deeds in the future'. This viewpoint rests on the assumption that the intended goals of stem cell research can be reached by using stem cell lines from spare embryos obtained in the private sector or abroad or from the existing stem cell lines that meet the standards imposed by the national law or regulations. They predict that the exclusion of new cell lines from spare embryos or stem cell lines derived from embryos solely created for the purpose of the research will not have significant negative impact on the progress of research.[44]

Those who think the separation doesn't hold

But let's face squarely the possibility of a failure in separation. Several arguments have been advanced to show that the separation between the use and derivation of ES cells cannot be guaranteed, and thus that research on cell lines already established by destroying human embryos does not avoid moral complicity in the killing of those who some consider as 'one of us'.

One of the most important objections to the policy decisions of the US and Germany is that a policy limitation to 'already existing cell lines' is an arbitrary line which may not hold in practice.[45] If ES cells do show clinical promise, and American and German researchers fall behind because of restrictions on their access to the cell lines that meet the required criteria, political leaders will come under pressure from researchers, (potential) patients and other stakeholders in ES cell research to change the regulations. The argument is that once you accept that there is or may be a great health benefit of the citizens, it is hardly possible not to extend the list of available stem cell lines when the existing ones appear to be an insufficient source, particularly where spare embryos are continually created as part of ART.[46] In 2003, eleven house republicans expressed their concerns about the quality, longevity, availability and terms of use of the stem cell lines in the NIH's ES cell registry and asked the president to review the 2001 policy for allowing for the creation of new ES cells lines.[47] Ruth Faden and John Gearhart (who was the first to derive and culture embryonic germ cells in 1998) have stated that the ES cell lines approved by Bush are inadequate to advance stem cell science. There are too few of them to accommodate the genetic diversity in our population and all were prepared using mouse cells, which may rule them out for clinical trials because of the risk that an animal virus might be passed to patients. Moreover, it is not possible to use federal funding to generate or study stem cells derived from embryos with genetic defects or disease genes. Such lines would be invaluable in gaining knowledge about the molecular basis of disease and in seeking ways to prevent or treat them.[48] Two recent studies shed light on other restrictions of the approved cell lines; The first, led by Fred Gage of the Salk Institute in La Jolla, California, and Ajit Varki of the University of California at San Diego, shows that the approved ES cell lines share a previously unrecognised trait that fosters rejection by the immune systems, diminishing their potential as medical treatments. A second study, led by Carol Ware of the University of Washington, is comparing characteristics of 14 of the 22 Bush-approved cell lines. At least five will never be useful in a clinical context because they are so difficult to grow; moreover, each colony has a tendency to turn into one kind of body cell or another, suggesting more than the 22 colonies will be needed if the field is to reach its full potential.[49]

Given the restricted usefulness of the approved cell lines, the question arises whether the US and Germany will allow research on other cell lines that do have the right characteristics to reach the intended research goals – which would be in accordance with the subsidiarity principle. But what if these lines appear insufficient, will they allow more ES cell research – involving the derivation of the ES cells if this is deemed necessary to reach the research goals?

Another problem with separation is that those who use the ES cells lend support and encourage those who derive the cells because they pay for these cells. According to Alex Capron, Director of Ethics, Trade, Human Rights and Health Law at the World Health Organisation, the distinction between paying for the *use* of stem cells and paying for their *derivation* is merely 'bookkeeping fiction'. The funding provided for studies using ES cells would flow directly to researchers deriving those cells, perhaps even in an adjacent lab. The only true difference would be that the federal funds would not go directly as a salary and laboratory expenses for the derivation process but indirectly as funds to purchase ES cells, which fund would then pay salaries, lab expenses, and so forth.[50]

No separation, but do we need a separation?

We can conclude that there is no real separation between the use of ES cells and their derivation. Given the lack of separation, support for research using ES cells is surely acceptable only to the extent that the process is acceptable.

In most countries with a stem cell policy based on the use-derivation distinction, the derivation of stem cells from foetal primordial germ cells is permitted. How can one explain the willingness to allow dead foetuses to be used but not discarded IVF embryos, when in most contexts the foetus, being further developed, is accorded a higher moral status? In Germany, foetal tissue research is not regulated by any special legislation. It is governed by regulations of the National Chamber of Doctors, and is federally and privately supported. Abortion is technically illegal in Germany, but women are not penalised, provided they receive counselling at a state-approved centre, which may then issue them a certificate. Moreover, it is astonishing that a government that wants to protect the embryo and its human dignity make the so-called 'abortion pill' RU-486 available.[51] Even more incompatible with strict views on embryo protection is that 10,000 children are born each year using IVF techniques and no stigma is attached to children so born, even though these were developed through research on embryos. This is because the respect most people now have for the early in vitro embryo is already low. Many people have concern for some kind of protection for embryos, but these feelings can change and depend on whether an embryo is involved in a parental project. IVF is a broadly accepted practice and has become routine in most countries. It is a practice in which thousands of spare embryos are intentionally created that will be discarded or will be used in medical experiments. In the US alone more than 400,000 embryos are stored in freezers awaiting their destination. Once people accept the creation and sacrifice of embryos to benefit infertile people, it seems inconsistent to condemn the creation and sacrifice of embryos to benefit those whose lives might be saved by stem cell therapies.[52] Moreover, it is important to point out that the IVF techniques of, cryopreservation, intracytoplasmic sperm injection (ICSI) and other techniques were all developed through research on embryos that came into being only for the purposes of the experiment. Some governments consider this type of experiment to be unacceptable from an ethical standpoint, although the results of such experiments are applied without any qualms and in most countries have even become routine. The same is true for embryo experiments that are currently done to develop methods to improve, facilitate or make reproduction possible, such as the development of better methods of in vitro culture and IVF, and of gamete and embryo storage.[53]

Many embryos are created and destroyed for no good purpose, because, for example, the mother did not think about the possibility of pregnancy and then decided that pregnancy and childbirth were too awful a prospect to contemplate. Equally, many embryos are created and destroyed for a good purpose – as part of a procreative endeavour that is completed or abandoned. These 'spare' embryos are then often destroyed, the good purpose for which they were created having been accomplished. Given that many thousands indeed many hundreds of thousands of embryos have been and will continue to be created to help to establish future lives should others (or indeed these embryos themselves) not be created to help save present and future lives? Why then might it be wrong deliberately to create an embryo not for itself but for the good using its tissue and cells might do? Grant the legitimacy of creating embryos that will perish as part of assisted reproduction, something that inevitably happens in all ART and in all countries that permit ART, then consistency surely demands permitting embryo sacrifice for morally equivalent purposes.[54] Even in Italy where current law requires the implantation of all embryos created in ART[55] it is recognised that such a law is effectively unenforceable because of the 'impossibility' for forcibly inserting embryos into unwilling and resisting women. Moreover as we shall shortly see, these problems of consistency apply as much to normal reproduction as they do to assisted reproduction.

There's evil and then there's 'evil'

It seems likely that if therapies were to be proved for serious illnesses using embryo derived stem cells, the people of Germany and France, countries which have condemned such research, would want to use them and their governments would find it difficult, if not impossible, to say 'no'. Even if the governments did refuse to accept effective drugs developed from ES cells, the Treaty of Rome, guaranteeing free movement in the European Union, would permit free movement of citizens for 'therapeutic tourism' to access therapies from neighbouring states. Equally, it would be unlikely (although logically consistent) for such governments to criminalise attempts to access such therapies as they perhaps do (or would be justified in doing) for those of their citizens who attempted to access, say, child sex in Asia contrary to the laws in their home European state.

How would we feel about the citizens of countries which have declared research on embryos unethical and have outlawed such research? Would they be hypocrites or monsters, or both? Should they so regard themselves? We think we would and should feel differently about a country which, while officially believing and saying that using embryo-derived stem cells is evil and hence that to use benefits derived from such cells would be benefiting from evil nonetheless accepted such benefits from other sorts of benefiting from evil.

Does this mean that we feel that there are evils and evils – different levels of evil? Or does it show that we do not actually believe that certain things claimed to be evil are in fact evil properly so called?

There are many possible, but two plausible, explanations for intuitive reactions here. The first is that there are indeed degrees of evil. Many believe adamantly that we should not accept any conceivable benefits flowing from the Nazi atrocities before and during the Second World War. For example the notorious 'Dachau Hypothermia Study' in which concentration camp inmates were subjected to lingering death in freezing water allegedly to study the survival possibilities for German pilots who crashed into the freezing North Sea or Atlantic Ocean. This 'science' has been condemned on many grounds, not least for the fact that it was poor science as well as unspeakable cruelty. However, opinions have differed about the ethics of using any real knowledge gained from such bestiality.[56] Those who think that the embryo is really 'one of us' and that experimenting on 'innocent' embryos would be (is in fact) like experimenting on innocent Jews, should however surely think there is a strong link between the two activities. There are two ways of resolving this. Either one could claim that there is in fact nothing wrong with benefiting from evil provided that the benefit is not responsible for the commission of the evil or for the commission of future analogous evils. That might be very difficult to establish in fact, although the principles can be articulated clearly. While there may be much to be said for this approach it is unlikely to satisfy those who find the Nazi atrocities so repugnant that the prospect of any good coming from them is simply unacceptable.

The alternative is to suggest that although people have said that embryo research is an evil, even those who condemn it and accept the embryo as 'one of us' do not in fact really believe any such thing.

Does anyone really believe that embryos are moral persons?[57]

Stem cell research and therapy using human embryos, and indeed all other therapeutic or research uses of embryos, might be successfully defended by drawing a distinction between what people say and what they do, or rather to point out that there may be an inconsistency between the beliefs and values of people as revealed by their statements, on the one hand, and by the way they behave on the other. 'To know the good is to do the good and to know the bad is to avoid the bad' and while this ancient 'truth' of course has to allow for weakness of will it does tell us something about consistency and sincerity. Although many people including most so called 'pro-life' or 'right-to-life' supporters are prone to make encouraging noises about the moral importance of embryos, and even sometimes talk as if embryos have, and must be accorded, the same moral status as you and me; they very seldom, if ever, behave as if they remotely believed any such thing. Taking for the moment as unproblematic the idea, made famous by Socrates, that 'to know the good is to do the good'; many Catholics, pro-life advocates or others who believe that the embryo is in a real sense 'one of us' don't behave consistently with their professed beliefs about what is good.

The embryo is not treated as 'one of us'

One would expect that those who give full moral status to the embryo, who regard it as a person like us, would both protect embryos with the same energy and conviction as they would their fellow adults and would mourn its loss with equal solemnity and concern. This however they do not do. It is true that some extreme defenders of the embryo in the United States have taken to murdering obstetricians who perform abortions, but those same individuals are almost always inconsistent in some or all of the following ways.[58]

We know that for every live birth up to five embryos die in early miscarriages.[59] Although this fact is widely known and represents massive carnage, pro-life groups have not been active in campaigning for medical research to stem the tide of this terrible slaughter. The loss of these 'spare' embryos is an inevitable and perhaps necessary part of normal sexual reproduction. Equally we know that for the same reasons the menstrual flow of sexually active women will often contain embryos. Funeral rights are not usually routinely performed over sanitary towels although they often contain embryos. In the case of spare embryos created by Assisted Reproductive Technologies, there has not been the creation of a group of pro-life women, offering their uteruses as homes for these surplus embryos. Indeed, some time ago I had to invent a fictitious quasi-religious order of women 'The Sisters of the Embryo' who would stand ready to offer a gestating uterus to receive unwanted embryos because (surprisingly given the large numbers of pro-life women available) there has never been such a movement, although there are records of isolated offers of wombs and of support for such a movement.[60] Indeed anyone engaging in unprotected intercourse runs substantial risk of creating an embryo that must die, and yet few people think that this fact affords them a reason either to refrain from unprotected intercourse (it is more usually the fear of creating an embryo that will not die that motivates them) or to press for medical research to prevent this tragic waste of human life.[61]

It is notorious that many would-be protectors of the embryo are prepared to permit abortions in exceptional circumstances, for example, to save the life of the mother or in the case of rape. However, in the former case the right course of action for those who believe the embryo has full moral status is to give an equal chance to the embryo or the mother (perhaps by tossing a coin) in cases where one may survive but not both. In the case of rape, since the embryo is innocent of the crime and has therefore done nothing to compromise its moral status, the permitting of abortion by those who give full status to the embryo is simply incoherent.[62]

These phenomena provide reasons for thinking that even if the views of those who believe the embryo to have the same moral status as normal adult human beings cannot be conclusively shown to be

fallacious, at least they can be shown to be inconsistent with the practice of most of those who profess such views, and that the 'theory' is
not therefore really believed by those who profess it or if believed is
actually compatible with the lives that human beings must, of necessity, lead. Stem cell research, including such research using human ES
cells, is then ethical and because of its immense promise, probably also
obligatory.

Notes

1 The authors acknowledge the stimulus and support of two European
Commission-funded projects. They are 'The Ethics of Stem Cell Research
and Therapy in Europe (Eurostem)', and the 'European Project on Delimit-
ing the Research Concept and the Research Activities (EU-RECA)', both
sponsored by the European Commission, DG-Research, 5th and 6th Framework respectively.

2 R. Lorenza (2004) 'Italian Minister in Trouble', *The Scientist* (September) 9.

3 A. A. Kiessling and S. Anderson (2003) *Human Embryonic Stem Cells. An
Introduction to the Science and Therapeutic Potential* (Boston, Massachusetts:
Jones and Bartlett Publishers), p. 164; D. Solter, D. Beyleveld, M. B. Friele et
al. (2003) *Embryo Research in Pluralistic Europe* (Berlin and Heidelberg:
Springer Verlag).

4 J. A. Thomson, et al. (1998) 'Embryonic Stem Cell Lines Derived from
Human Blastocysts', *Science*, 282(5391), pp. 1145–7. R. A. Pedersen (1999)
'Embryonic Cells for Medicine', *Scientific American* 1280(4), pp. 68–73;
B. E. Reubinoff et al. (2000) 'Embryonic Stem Cell Lines from Human
Blastocysts: Somatic Differentiation in vitro', *Nature Biotechnology*, 18,
pp. 399–404. R. E. Schwartz et al. (2002) 'Multipotent Adult Progenitor Cells
from Bone Marrow Differentiate into Functional Hepatocyte-like Cells',
Journal of Clinical Investigations, 109(10), pp. 1291–302; C. M. Verfaillie
(2002) 'Adult Stem Cells: Assessing the Case for Pluripotency', *Trends in Cell
Biology*, 12(11), pp. 502–8. Austin Smith has emphasised the importance of
pursuing research on all sources of stem cells simultaneously (paper presented at FENS Forum Workshop, Paris, 13 July 2002). G. Kogler et al.
(2004) 'A New Human Somatic Stem Cell from Placental Cord Blood with
Intrinsic Pluripotent Differentiation Potential', *Journal of Exploratory
Medicine*, 200(2), pp. 123–35.

5 F. D. Camargo, S. M. Chambers and M. A. Goodell (2004) 'Stem Cell
Plasticity: From Transdifferentiation to Macrophage Fusion', *Cell Proliferation*, 37, pp. 55–65; C. Mummery (2004) 'Stem Cell Research: Immortality
or a Healthy Old Age?', *European Journal of Endocrinology*, 151 (November),
Suppl. 3, pp. U7–U12.

6 See, for example, J. R. Meyer, a priest of the Opus Dei Prelature, who says
that 'the medical benefits which might accrue for some patients do not outweigh the grave consequences for the embryo that is killed in order to
produce ES cells for medical therapy'. J. R. Meyer (2000) 'Human Embryonic
Stem Cell and Respect for Life', *Journal of Medical Ethics*, 26, pp. 166–70.

7   Maurizio Mori gives an account of the origin of the expression 'the embryo, one of us': The story is the following. It was used by Professor Francesco D'Agostino, Chair of the Italian National Ethics Committee when he presented the result of the Report of the Italian National Committee for Bioethics to the media in 1996. He said more precisely that the Committee had stated 'the human embryo is to be treated as one of us'. Of course it was immediately shortened to the more famous form. There is, however, at least one book with such a title: Gino Concetti (1997) *L'embrione uno di noi* (Rome: Vivere In). The phrase 'the embryo is one of us' is now associated with the Pope John Paul II and is widely used in pro-life discourse. See Congregation for the Doctrine of the Faith. Instruction on Respect for Human Life in its Origin and on the Dignity of Procreation, Replies to certain Questions of the Day, 22 February 1987. At: http://www.vatican.va/roman_curia/congregations/cfaith/documents/rc_con_cfaith_doc_19870222_respect-for-human-life_en.html. And by the Canadian Conference of Catholic Bishops on Bill C-13 An Act Respecting Assisted Human Reproduction. 19 January 2003. At: http://www.cccb.ca/PublicStatements.htm?CD=343&ID=1258.

8   S. Holm (2002) 'Going to the Roots of the Stem Cell Controversy', *Bioethics*, 16(6), pp. 493–507.

9   Health Council of the Netherlands (2002) *Stem Cells for Tissue Repair: Research on Therapy* (The Hague, 27 June), p. 46. G. Pennings and A. Van Steirteghem (2004) 'The Subsidiarity Principle in the Context of Embryonic Stem Cell Research', *Human Reproduction*, 19(5) (May), pp. 1060–4. The European Group on Ethics in Science and New Technologies to the European Commission (2000) *Adoption of an Opinion on Ethical Aspects of Human Stem Cell Research and Use* (Paris: European Commission, 14 November) (Opinion N° 15), p. 14.

10   National Bioethics Advisory Commission (1999) *Ethical Issues in Human Stem Cell Research* (Rockville, MD: NBAC, September), p. 53.

11   German Bundestag. Act ensuring protection of embryos in connection with the importation and utilization of human embryonic stem cells. Stem Cell Act, s. 5. Berlin, 28 June 2002.

12   The validity of the principle in this context can be questioned for several reasons. First of all it is unclear whether this presupposed hierarchy is defensible. Second, the principle may be anti-ethical in the sense that it implies that we have to follow existing public opinion, whether this is well argued and well informed or not. Third, people make a different evaluation of the available scientific evidence, and also have a different approach to the decision as to whether a certain line of research should be deemed 'necessary'. Opponents of ES cell research claim that alternatives exist, which do not require the 'instrumental use' of human embryos. However, we can ask ourselves whether it isn't misleading to present every alternative, which does not use embryos, as *a priori* superior. For comparative ethical analysis a number of relevant aspects should be taken into account, including the burdens and risks of a certain method, the chances that the alternative options have the same applicability as ES cells, and the costs and the time scale in which useful clinical applications are to be expected. Finally, it is said to be a reasonable principle in policy decisions, but is it as

reasonable as is said when we know that it leads to delay in the development of treatments that can alleviate the suffering of thousands, maybe millions of people? Holm, 'Going to the Roots of the Stem Cell Controversy'; P. A. Roche and M. A. Grodin (2000) 'The Ethical Challenge of Stem Cell Research', *Women's Health Issues*, 10(3), pp. 136–9.

13 R. Lovell-Badge (2001) 'The Future for Stem Cell Research', *Nature*, 414(6859), pp. 88–91; C. R. Cogle et al. (2003) 'An Overview of Stem Cell Research and Regulatory Issues', *Mayo Clinic Proceedings*, 78, pp. 993–1003.

14 B. E. Edwards, J. D. Gearhart and E. E. Wallach (2000) 'The Human Pluripotent Stem Cell: Impact on Medicine and Society', *Fertility and Sterility*, 74(1), pp. 1–7.

15 In the curious case of the USA.

16 R. Doerflinger (1999) 'The Ethics of Funding Embryonic Stem Cell Research: a Catholic Viewpoint', *Kennedy Institute of Ethics Journal*, 9(2), pp. 137–50; W. Friend (2003) 'Catholic Perspectives and Stem-cell Research and Use', *Origins*, 32(41), pp. 682–6; J. Oakley (2002) 'Democracy, Embryonic Stem Cell Research, and the Roman Catholic Church', *Journal of Medical Ethics*, 28(4), p. 228.

17 Belgian Senate. Belgian Law Concerning Research on the Embryo In Vitro. 11 May 2003. Human Fertilisation and Embryology Authority. HFEA Act 1990 extended by the Human Fertilisation and Embryology Regulation of 2001. In November 2004, the Swedish government Bill 2003/04:148 on stem cell research was enacted. The Bill explicitly allows the creation of embryos through somatic cell nuclear transfer and will came into force on 1 January 2005. The Dutch Embryo Act of 2002 prohibits the creation of human embryos for research. However, this ban is not irreversible and could be lifted by Royal Decree within five years of the Act coming into effect.

18 These terms are used in R. M. Green (2001) *The Human Embryo Research Debates: Bioethics in the Vortex of Controversy* (Oxford and New York: Oxford University Press).

19 For an overview of these legislations, see C. M. Romeo-Casabona (2002) 'Embryonic Stem Cell Research and Therapy: the Need for a Common European Legal Framework', *Bioethics*, 16(6), pp. 557–67.

20 R. Lorenzi (2003) 'Italy Approves Embryo Law', *The Scientist*, 12 December.

21 Article 40.3.3 of the Irish Constitution expressly prohibits research on embryos. It acknowledges the right to life of the unborn and states that it is equal to that of the mother.

22 However, the import of existing ES cell lines is not prohibited. Most members of the Bioethics Commission are in favour of allowing research on imported ES cell lines created before a set date. Bioethics Commission at the Federal Chancellery of 3 April and 8 May 2002. Stem cell research in the context of the EU's Sixth Framework Programme Research. At: http://www.austria.gv.at/2004/4/18/stem_cell_research.pdf.

23 The Norwegian Biotechnology Advisory Board. Biotechnology Act 1994. At: http://www.bion.no/lov/lov-19940805-056-eng.pdf.

24 K. Wheat and K. Matthews, *World Human Cloning Policies*. http://www. ruf.rice.edu/~neal/stemcell/World.pdf.

25 G. Pennings (2002) 'Reproductive Tourism as Moral Pluralism in Motion', *Journal of Medical Ethics*, 28, pp. 337–41, and (2004) 'Legal Harmonization and Reproductive Tourism in Europe', *Human Reproduction*, 19(12), pp. 2689–94.

26 S. Arie (2004) 'Italians Force Referendum on Fertility Law', *Guardian Unlimited*, 1 October. At: http://www.guardian.co.uk/italy/story/0,12576, 1317232,00.html.

27 A. Haverty (2003) 'Ireland Divided on Stem Cells', *The Scientist*, 26 November.

28 See, for example, the National Institutes of Health (2000) *Guidelines for Research Using Human Pluripotent Stem Cells* (Bethesda, MD: NIH). German National Ethics Council (2001) *Opinion on the Import of Human Embryonic Stem Cells* (Berlin: German National Ethics Council) (under the heading 'specific arguments in favour of the import of human embryonic stem cells').

29 Letter from HHS Gen. Counsel Harriet Rabb to Harold Varmus, at that time Director of the NIH on 'Federal Funding for Research Involving Human Pluripotent Stem Cells', 15 January 1999.

30 G. J. Boer (1999) 'Ethical Issues in Neurografting of Human Embryonic Cells', *Theoretical Medical Bioethics*, 20(5), pp. 461–75. G. Pennings (2003) 'Ethical Issues Regarding Embryonic Stem Cells', in *Lectures in Medicine: Embryonic Stem Cells*, organised by the Belgian Faculties of Medicine (Brussels: AZ-VUB, 6 February).

31 Opponents of the use of foetal cells, or tissues obtained following clinical abortion, claim that all those who isolate and use the foetal material are accomplices in the preceding abortion and that it will lead to an increase of the number of abortions. Advocates of foetal tissue research claimed that these objections could be bypassed by guaranteeing a separation between the act of abortion and the use of foetal material for research and therapies. Many countries have tried to guarantee this separation by imposing conditions on the performance of abortions and on the donation of fetal tissue in laws and regulations. A. F. Shorr (1994) 'Abortion and Fetal Tissue Research – Some Ethical Concerns', *Fetal Diagnosis Therapy*, 9(3), pp. 196–203; J. C. Rankin (1990) 'The Fetal Tissue Debate on Complicity', *Hastings Centre Reports*, 20(2), p. 50.

32 R. K. Zimmerman (2004) 'Ethical Analyses of Vaccines Grown in Human Cell Strains Derived from Abortion: Arguments and Internet Search', *Vaccine*, 22, pp. 4238–44.

33 American Association for the Advancement of Science and Institute for Civil Society (1999) *Stem Cell Research and Applications: Monitoring the Frontiers of Biomedical Research* (Washington, DC, November), p. 9.

34 Green, *The Human Embryo Research Debates*.

35 German National Ethics Council (2001).

36 Green, *The Human Embryo Research Debates*, p. 554.

37 Green, *The Human Embryo Research Debates*, p. 552.

38 R. A. Charo (2001) 'Bush's Stem Cell Compromise: a Few Mirrors?' *Hastings Centre Reports*, 31(6), pp. 6–7.

39 See, for example, Option A and B of the German National Ethics Council (2001).

40 A. M. Capron (1999) 'Good Intentions', *Hastings Centre Reports*, 29(2), pp. 26–7.
41 The White House (2001) Office of the Press Secretary. Radio address by the President to the nation. Bush Ranch, Texas. 11 August. At http://www. whitehouse.gov/news/releases/2001/08/20010809-2.html. For the eligibility criteria, see http://stemcells.nih.gov/registry/eligibilityCriteria.asp.
42 Bush, radio address by the President (2001).
43 'Provided that their import has no causal effect on the production of embryonic stem cells abroad, the use of embryos for the production of stem cell lines cannot be deemed to be the responsibility of the research workers or of the German State authority.' German National Ethics Council (2001).
44 AAAS and ICS, *Stem Cell Research and Applications*, p. 18.
45 US Catholic Bishops. Statement on Cloning and Embryo Research. 13 August 2001. At http://www.usccb.org/prolife/issues/bioethic/fact801.htm.
46 H. Gottweis (2002) 'Stem Cell Policies in the United States and in Germany: Between Bioethics and Regulation', *Policy Studies Journal*, 30(4), pp. 444–69.
47 M. Castle et al. (2003) 'Letter to President Bush on Stem cell Research from 11 House Republicans', 15 May. Http://www.aaas.org/spp/cstc/briefs/stem-cells/stemhsltr.shtml.
48 R. Faden and J. Gearhart (2004) 'Facts on Stem Cells', *Washington Post*, 23 August, p. A15.
49 R. Weiss (2004) 'Approved Stem Cells' Potential Questioned', *Washington Post*, 29 October, p. A03.
50 A. M. Capron (2002) 'Human Embryonic Stem Cell Research: Ethics and Politics in Science Policy', in Shui Chuen Lee (ed.) *Proceedings of the Third International Conference of Bioethics* (University of Chungli, R.O.C. Taiwan, June), p. V–12.
51 J. Harris (2003) 'Stem Cells, Sex and Procreation', *Cambridge Quarterly of Health Ethics*, 12(4), pp. 353–72.
52 For a critical analysis of the moral difference between the use of spare IVF embryos for research and the use of embryos solely created for the purpose of research, see K. Devolder (in press 2005), 'Creating and Sacrificing Embryos for Stem Cells', *JME*.
53 Solter et al., *Embryo Research in Pluralistic Europe*.
54 See John Harris (2003) 'Stem Cells, Sex and Procreation', *Cambridge Quarterly of Healthcare Ethics*, 12(4) (Fall), pp. 353–72.
55 On 19 February 2004 the President of the Republic of Italy promulgated Law 40/2004 'Norms on the matter of medically assisted procreation' [Legge 40 Norme in materia di procreazione medicalmente assistita] (Republic of Italy, 2004). http://staminali.aduc.it/php_leggi.html.
56 A. Caplan (ed.) (1992) *When Medicine Went Mad: Bioethics and the Holocaust.* (Totowa, NJ: Humana Press); P. Hoedeman (1991) *Hitler or Hippocrates: Medical Experiments and Euthanasia in the Third Reich* (Sussex: Book Guild); B. Muller-Hill (1988) *Murderous Science: Elimination by Scientific Selection of Jews, Gypsies, and Others in Germany, 1933–1945* (Oxford: Oxford University Press); R. J. Lifton (1986) *The Nazi Doctors: Medical Killing and the Psychology of Genocide* (New York: Basic Books).
57 In this section we have benefited from discussions with Dan Wikler. See also John Harris (2003) 'Stem Cells, Sex and Procreation', *Cambridge*

Quarterly of Healthcare Ethics, 12(4) (Fall), pp. 353–72; and (2004) 'The Great Debates – Julian Savulescu and John Harris', *Cambridge Quarterly of Healthcare Ethics*, 13(1) (January), pp. 68–96. John Harris's contributions to this debate: 'Sexual Reproduction is a Survival Lottery', pp. 75–90, and Julian Savulescu and John Harris (2004) 'The Creation Lottery: Final Lessons from Natural Reproduction: Why Those Who Accept Natural Reproduction Should Accept Cloning and Other Frankenstein Reproductive Technologies', *Cambridge Quarterly of Healthcare Ethics*, 13(1) (January), pp. 90–6.

58 It is difficult to avoid the conclusion that murder is more congenial to them than more strenuous but less satisfying ways of preventing deaths of embryos.

59 See J. Harris (2003) 'Stem Cells, Sex and Procreation', *Cambridge Quarterly of Healthcare Ethics*, 12(4) (Fall), pp. 353–72.

60 J. Harris (1992) *Wonderwoman & Superman: The Ethics of Human Biotechnology* (Oxford: Oxford University Press), p. 47.

61 This point has been developed in some detail in J. Harris (2002) 'The Use of Human Embryonic Stem Cells in Research and Therapy', in Justine C. Burley and John Harris (eds.) *A Companion to Genethics: Philosophy and the Genetic Revolution* (Oxford: Basil Blackwell), pp. 158–75; and in J. Harris (in press) 'Stem Cells, Sex and Procreation', *Cambridge Quarterly of Healthcare Ethics*.

62 J. R. Richards (1982) made this point well in *The Sceptical Feminist* (Harmondsworth: Penguin Books). Note also the similarities between this argument and Athanassoulis's point on the sanctity of life in Chapter 8 in this volume.

5
Towards a Natural Law Critique of Genetic Engineering

David S. Oderberg

Introduction

One of the major factors affecting contemporary academic debate about biotechnology in general, and genetic engineering in particular, is that virtually all participants approach the issues from the standpoint of consequentialist moral theory. Now consequentialism takes many forms, especially when it comes to the question of just what it is that rational agents should be seeking to maximise. We can, however, abstract from these variations and note that as far as biotechnology is concerned, all consequentialists frame the terms of the debate around the key question of what harms or benefits a given technique, practice or method is likely to bring about.

One result of this phenomenon is that bioethical debate is skewed in favour of an ethical theory that is on many counts implausible at best, positively dangerous at worst.[1] Another is that alternative ethical approaches typically receive short shrift. Rights-based theories (to the extent that rights *on their own* can form the basis of an ethical theory, which is doubtful) do receive some attention. On the other hand, natural law theory (NLT), to the extent that it receives any attention, is invariably brought into the discussion primarily for the purposes of exposing its many alleged fallacies and then consigning it to the theoretical dustbin. There is perhaps one positive aspect to this treatment, namely that opponents of NLT have emphasised the implausibility or irrelevance of a number of superficial concepts and distinctions that do not support a critique of biotechnological practices. That the opponents rarely if ever try to reach beyond these superficialities, however, means that an ethicist unfamiliar with NLT will almost certainly come to think of it as little more than a congeries of vague ideas and blatant non-sequiturs. Since it is those with a reflex hostility to NLT who also explain the theory to each other and to outsiders, and who define the terms of the debate and provide the criticism, it is no wonder that NLT is held in such low regard in bioethics.

Natural law theorists are, therefore, obliged to show how their approach to ethical questions does, contrary to the many critics, provide a reasoned and reasonable response to the issues raised by biotechnology. The obligation is all the more pressing because of the fraught nature of these issues, which lie at the heart of public policy and are the subject of much anxiety outside the narrow realm of academic ethics. In the contribution to this enterprise that follows, I begin first with an account of how NLT should *not* be understood. The essay then moves on to an exposition of the positive case for natural law as applied to bioethics in general, followed by its application to some major questions of genetic engineering. Central to my outline and defence of NLT in the realm of bioethics will be an attempt to spell out the proper distinction between the natural and the unnatural that lies at the heart of the theory. It is by seeing how that distinction is *not* to be understood that we can begin to gain an idea of what the natural law theorist *does* in fact claim about the relationship between ethics and nature.

How not to understand the natural/unnatural distinction

A cluster of criticisms levelled by bioethicists at NL theorists centre on the alleged distinction between the 'normative' and the 'descriptive'. The used of scare quotes is advisable, since moral theorists vary about (i) what these terms mean, (ii) what the distinction is claimed to amount to, and (iii) what kind of mistake NL theorists are supposed to be making when, ostensibly, they invoke it.

Untouched nature

Peter Singer and Deane Wells, in their attack on NL theory,[2] refer to the 'simple-minded version' of the 'unnaturalness' objection to IVF, but the point applies to other controversial biotechnological proce- dures, whereby 'what occurs in nature untouched by human interven- tion' can be a guide to how we should act. This 'descriptive view', as they call it, would require the rejection of all medical treatment and of any other intervention in 'the world apart from human beings'.[3] Needless to say, they give no example of an NL theorist who espouses such a view, nor can one be found (inside or outside academia, I imagine). Nor should any NL theorist give the slightest credence to the idea that one can 'read off' from what happens to occur in nature the truth about how we should behave. NL theory should not and does not imply that somatic cell nuclear transfer is wrong because it does not occur in nature, any more than that setting broken legs is wrong because it does not happen in nature. Ethics is not a branch of zoology, biology or even anthropology (though all of these inform ethical reasoning in various ways). Nor does NL theory presuppose the 'notion that the normal or natural course of events is a benign one, and hence we ought not to seek to alter it'.[4] Nature can indeed be cruel, and mankind has throughout history intervened in nature to prevent and cure disease and to make physical existence as comfortable as possible. Indeed, NL theory can, and NL theorists often do, take this very fact into account when reasoning about how we should act.

It is, however, a step too far for Singer and Wells to say that '[n]ature – the course of events that we do not control – is blind to the welfare of any creature'.[5] Nature is not merely a series of floods and famines, but a complex network of laws and processes that enable the sort of adaptation to their environment that makes animal (and human) survival – in most situations, most of the time – possible in the first place.[6] As we shall see, a proper understanding of NLT recognises that nature must be very much our guide, though not our master, in the appreciation of what makes life go well. So we can quickly dispense with this straw man raised by Singer and Wells. What is 'monstrous', to use their epithet, is less 'to believe that we should leave the world alone, meekly suffering the consequences without daring to challenge the "natural" course of events',[7] but to believe that any ethicist – let alone any natural law theorist – ever held such a view in the first place.

The normal course of events

Another interpretation of the natural/unnatural distinction hinges on whether a given practice is *normal*, where this is defined, in the words of Singer and Wells, as 'occurring in the normal course of events'.[8] Needless to say, the proposed definition is circular unless there is a further explanation of what the normal course of events is supposed to amount to. Although the authors are not explicit, it seems they construe it as a statistical or probabilistic phenomenon: what occurs in the normal course of events is what most commonly happens, or at least what is most likely to happen, in a given situation. To use their example, in the normal course of events people die from advanced cancer; but this does not imply that one should refuse to depart from such a course of events by seeking to apply potentially successful new treatments.

Again, any NL theorist who supposed that the natural/unnatural distinction could be founded on a statistical or probabilistic account of nature would be mistaken, and needless to say no theorist, as far as I am aware, has done so. Nevertheless, one perhaps hears non-philosophers talk in this way, for instance when people speak of 'letting nature take its course'. But even here, we should be charitable in our interpretation of non-philosophical ways of speaking. It is, I believe, fair to suppose that what most people mean, on reflection, by letting nature take its course is something subtler and more complex than that the statistically normal course of events should not be resisted; moreover, what someone means by such a locution will depend on the context. The idea of letting nature take its course encompasses a range of thoughts, concerning such matters as: the distinction between using ordinary and extraordinary means to keep dying people alive; the undesirability of interfering in physical processes of which we are substantially ignorant; the potential harm in undermining, even with the best of intentions, the self-healing and homeostatic powers intrinsically possessed by organisms; and so on. These sorts of idea are at the heart of a more profound and plausible understanding of the natural/ unnatural distinction, as we shall see. The role of statistics and probability is not to provide *independent* grounds for appealing to the natural as against certain kinds of biotechnology. Rather, it is to provide *evidence* pointing to fundamental characteristics of human life and biological processes in general that ground ethical judgement.

What God intends

Singer and Wells also consider the possibility that an ethical appeal to the unnatural is an appeal to what God intended.[9] This they dismiss for epistemological reasons, as well as for the allegedly absurd consequence of His allowing the invention and development of practices that are supposedly contrary to His will. The first objection is of little force in itself, since if a NL theorist *were* to appeal to God's intent, he would of course be bound to show why his method for determining that intent was more accurate than some other method yielding contradictory moral conclusions (e.g. a different source of revelation, or a different theological tradition). That there is wide disagreement among theists about how to determine God's intent, and indeed about religion in general, is no more an obstacle in itself to reaching a right answer than wide disagreement in ethics is, contrary to the relativist charge, an obstacle to reaching ethical truth.[10] The ethicist who appeals to God's will – whether he be a NL theorist or not – has to make out a rational case just like anyone else, and cannot be ignored simply because he has, like everyone else, to give an account of the sources of moral knowledge and the methods of reaching it.

As to the second charge, Singer and Wells have little to say in support other than that if the standard theological reply were true – namely, that God allows the existence of practices of which He disapproves because He allows human beings to exercise their freedom to do wrong – then this would be 'a very unflattering account of Divine behaviour'[11] since God, being omniscient, would already know the results of His test of our free will. One would not have thought, on reading their account, that anyone had ever tried, let alone succeeded, in reconciling free will and God's foreknowledge. Moreover, one would have expected that Singer and Wells, having criticised the idea that what occurs in nature is *ipso facto* good, should have welcomed the more plausible idea that there are things that occur in nature but are not good. For as far as moral philosophy is concerned, this is what the appeal to God's intent amounts to; *why* such things occur takes one into the realm of theology, but *that* they occur is something all ethical theorists can agree on.

Moreover, once again the critics of natural law, of which Singer and Wells are typical examples, have misunderstood what the NL theorist claims. It is quite true, and not to be minimised, that the vast majority of NL theorists are theists of one stripe or another, mainly Christian. And this is no accident, since the very idea of natural law demands ontological support of the kind provided by theism. The NL theorist, as we will see, believes in a *natural order of things* that structures the world antecedently of human desires and preferences, and that morality concerns the way in which those desires and preferences interact with a prior metaphysical order. When coupled with the proposition, central to NL theory, that moral truths are in some sense necessary and immutable, an ontological foundation of the kind theism provides is an obvious, perhaps inevitable, corollary of the theory.

Nevertheless, NL theorists do *not*, at least typically, use appeal to divine sanction as the *epistemological* ground of moral truth. The whole point of being a *natural* law theorist as opposed, say, to a 'divine command' theorist, is that one believes that human beings can both discern the existence of a natural order without special appeal to the divine (though without being rational theists that discernment will be imperfect) and have access to moral truth through reflection upon that

natural order. Moreover, this view is compatible both with the existence of certain *specific* moral injunctions that can only be known
through, say, revelation, and with the NL theorist's position that, in
ultimate ontological terms, the natural law just *is* the eternal law
ordained by God. Hence the sorts of superficial criticism typically levelled at NL theorists over the role of God in their theory miss the mark.

We are part of nature

Singer and Wells ask: 'Are we not ourselves part of nature? In designing
ways of overcoming obstacles to the fulfilment of deep human desires,
are we not acting with our own rational nature?' And they cite with
approval the Protestant theologian and bioethicist Joseph Fletcher,
who claims that the proper ethical distinction is not between the
natural and the artificial, but between the random and the 'rationally
willed or chosen'.[12] It is true, as he adds, that having rationality and
freedom is what distinguishes humans from other animals. But it is no
part of NL theory that humans should simply do what distinguishes
them from other species without regard to the *proper objects* of the
rational will. NL theory does not emphasise rationality independently
of the objects at which rationality should be directed; hence there can
be no place for a kind of fetishism of choice of the sort one finds in
ethical liberalism and autonomy-based theories.

The NL theorist not only recognises that we are part of nature, but
places this fact at the centre of his moral perspective. The distinction
he wishes to draw between the natural and the unnatural is not the
morally specious one between what occurs in the world untouched by
human hands and the world as shaped and moulded by human action.
Rather, it is, as I will explain, between what does and does not fulfil the
natures of things in general and human nature in particular. Whether
we are concerned with the world of living things unaffected by human
choices or the world of living things – of humans in particular – as
shaped by human decision-making, we are able, so the NL theorist
argues, to make a distinction between what is natural and unnatural
according to how the phenomenon in question – be it a process, an
action, an event, a choice – fulfils the natures of the objects under
consideration.

A natural law account of genetic engineering

All of the superficial criticisms of natural law theory mentioned above
achieve no more than to make it look like a simplistic account of

morality meriting barely a nod of recognition. Yet they can be seen as deriving mainly from contemporary ethics' obsession with the supposed problem, as the parlance goes, of deriving an 'ought' from an 'is'. This is not the place for a detailed discussion of the so-called is-ought distinction and the supposed 'naturalistic fallacy'.[13] But it is crucial to a proper appreciation of NL theory that its attitude to the alleged problem be made clear.

The charge of committing the 'naturalistic fallacy' is well expressed by Michael Reiss and Roger Straughan: 'To assume that we can simply deduce what is morally right and wrong from certain facts about the world and about Nature is to commit what philosophers have called the "naturalistic fallacy" ... simply because something happens in nature does not mean that it is right or good, that it should be preserved or respected.'[14] The charge is a familiar one, yet familiarity does not breed justification. The very dichotomy on which it is based is rooted in the empiricist legacy, according to which the very idea of the normative in nature is incoherent. The essence of the naturalist programme in moral philosophy, as explained clearly by John Cottingham in his recent essay charting the idea of nature in the early modern period, has been 'to explain the realm of the normative (including the domain of moral obligation) in broadly empirical terms – as somehow part of, continuous with, or in some sense derivable from, the ordinary natural phenomenal world around us.'[15] By the very terms of the project, the natural phenomenal world was construed as the object of empirical observation divorced by definition from normative concepts. It doomed from the outset the natural law theorist's traditional conception of the natural, since for him the natural *embeds* normative phenomena within it – phenomena such as goodness and badness, fulfilment, proper functioning, essential integrity, flourishing, applied both to individuals and to society.

One might wonder whether, had the so-called Scientific Revolution occurred independently of empiricist influence (particularly from Bacon and later Locke), the very concept of empirical observation handed down to us would have been radically different, and more in tune with the pre-empiricist understanding with which NL theory operates. Historical musings aside, we must note for present purposes that the same empiricist, 'naturalistic' ideas have been passed on to contemporary ethics pretty much unchanged. They are at the heart of the charges listed earlier, and emerge in the various forms of consequentialism and pragmatism that dominate current bioethical debate. In looking at genetic engineering, the typical bioethicist sees it virtually as the geneticist himself sees it: a collection of techniques and practices that can be used for good or for ill, but that of themselves have no inherent ethical character. What ethical character we choose to give them – indeed, whether we choose to give them any at all rather than simply to regard them as morally neutral – depends on our individual preferences, the preferences of society, the uses to which they can be put and, most especially, whether they are likely to cause harm (or benefit).

At this point one might wonder what role harm plays in the anti-NLT approach to bioethics. After all, isn't harm a normative concept, and isn't the only way of finding out whether a practice causes harm to observe its effects empirically? If so, doesn't the critic of NLT commit the same 'descriptive-normative' sin of which he accuses his opponent? There are various ways in which the sorts of bioethicist I have in mind interpret the concept of harm. (For convenience, I will now call these theorists CP bioethicists, for 'consequentialist-pragmatist', whilst recognising that they take a variety of approaches, that consequentialism and pragmatism are not the same position even though they overlap substantially, that I am omitting consideration of rights theorists, and so on.) Some take harm to be intrinsically normative, as, say, giving us a *prima facie* reason not to do the thing that causes it. Others locate the normativity in our attitude to the activity in question, claiming that what is harmful depends primarily on whether it is something to which the person who is caused the harm is averse, or would be averse on rational reflection, and so on. But all CP bioethicists who take harm to be, in one sense or another, a factor in determining the morality of a biotechnological practice, are subject to the same criticism. This is that they themselves must recognise the existence of a *natural order of things*, according to which harm is in general bad for the one who is caused it. The point holds whether harm is taken to be intrinsically bad, or whether the badness derives from the attitude of aversion on the part of the victim.[16]

Take a simple question, such as why it is morally wrong for parents to neglect their children. The general reply given by a consequentialist is that children who are neglected suffer, and suffering is bad. But why is suffering bad? Why should it matter? If the consequentialist appeals to the thought that harm just is, intrinsically, a bad state of affairs, he is doing no more or less than agreeing with the NL theorist that in the natural order of things, harm is bad. (I leave aside further issues concerning whether all harm is bad, whether some harm is always bad, whether harming must be distinguished from wronging, and so on.[17])

If he appeals to the aversive attitude of a victim of harm, again he simply echoes the NL theorist that, in the natural order of things, harm tends to lead to aversion. The consequentialist will almost certainly entwine his normative theory with qualifications from liberalism concerning consent, self-inflicted harm, and the like, but he will still take it that in the paradigmatic, non-consensual case, harm – whether of itself or because of its effects – is a bad whose infliction can be justified only by further appeal to maximisation of the good. A world in which parental neglect did not tend to harm children might be conceivable, but it would not be *our* world, and for that reason it would not be natural since what is natural for human beings depends on what holds in our world, or at least in worlds very similar to our own.

Again, when it comes to preferences, there is a persistent tension in utilitarian claims about maximisation. Should preferences be taken at face value, or must they meet a standard of reasonableness?[18] To take them at face value would mean, for instance, that should the utilitarian 'calculus' favour the genetic engineering of new species in a way that caused immense pain and suffering to untold numbers of human beings (e.g. through experimentation), there could be no avoidance of the practical conclusion that the practice must go ahead. The results of such a policy in respect of preferences would be abhorrent by just about any standard. Yet if a reasonableness constraint were built into the criterion of preference satisfaction, morally repugnant results might be avoided (at least some of the time) but at the cost of recognising certain features of human life, such as the intrinsic badness of suffering, that the NL theorist has recognised all along.

For all the other theoretical differences between NLT and a consequentialist or pragmatist approach to ethics, it seems, then, that when it comes to bioethics there is common agreement over the centrality of the question whether a practice or technique is harmful. Yet, to echo Alasdair Macintyre, one might still ask: 'Whose harm? Which morality?' For the problem, as the NL theorist sees it, is that when it comes to applying the concept of harm, the CP bioethicist wilfully wears moral blinkers. For instance, Matti Häyry asserts that 'genuinely pragmatic considerations for and against gene technology are decisively important',[19] and this he glosses by reference to questions such as whether genetic engineering is dangerous.[20] As for natural law, he writes it off by appeal to just the sorts of superficial objections and distinctions mentioned above.[21] Ulla Wessels, to take another typical example, thinks that the ethics of genetic engineering is all about weighing risks and benefits,[22] where these do not go beyond such things as suffering (understood as physical pain)[23] and loss of human variety (whatever that means, exactly).[24] Reiss and Straughan, in their discussion, are mainly exercised by 'extrinsic concerns about risk and safety', which have to be weighed on a case-by-case basis in typical utilitarian fashion.[25]

For the natural law theorists, these recurrent concerns about genetic engineering, coming almost exclusively from those writers who generally support it, are based on a distorted conception of just what sort of harm a practice might tend to cause, which only a proper understanding of the natural/unnatural distinction can rectify. NLT holds that we need to look at the conditions under which human nature is fulfilled – what causes it to flourish, and what causes it to be perverted, damaged or impaired. Those practices which conduce to the fulfilment of human nature are natural, and those that do not are unnatural. The so-called 'naturalistic fallacy' and the alleged problems of arguing from the 'descriptive' to the 'normative', raised almost ritualistically by critics of NLT, are bypassed entirely. The distinction itself cannot be made out since the normative is built into the very description of nature in the first place. Hence there can be no place for naturalism in the way the heirs of empiricism define it, based as it is on a supposed norm-free description of nature, including human nature. On the other hand, the NL theorist is quite happy to be called a 'naturalist' in the sense of basing his moral theory squarely on the natural, where the account of the natural includes a proper understanding of what does and does not fulfil human nature. If that is what naturalism amounts to, then the NL theorist is a naturalist, only the meaning and implications of such a naturalism are very different from those deriving from the 'naturalism' bequeathed by empiricism.

The natural law theorist's criterion of the natural is usually spelled out in terms of a list of *basic goods* the pursuit of which fulfils human beings.[26] Although there is some disagreement over what is on this list, there is also a lot of overlap, far more, for what it is worth, than one would find in a comparison of consequentialists' views on the object of maximisation.[27] One thing, for instance, upon which virtually all natural law theorists would agree is that a practice whose effect was to undermine family life would be immoral. The good of family, says the NL theorist, is part of the good of friendship, and family life is key to the fulfilment of human nature. This does not mean, of course, that being a hermit or a recluse is *ipso facto* immoral, or that one must always get on with one's family, and so on: human goods come as a package, as it were, and different people are free to place different emphases on different goods in their own lives, provided that they respect and do not violate any good thereby.

If a practice, policy or state of affairs undermines family life, it is, on NLT, wrong or bad. Such might be the case, say, for excessive taxation of families, for a policy that made it difficult for a mother to choose freely to stay at home and look after her children, or for government intervention in the way children are brought up (all of these being subject to various qualifications, of course). It would also – so I would argue, and many NL theorists would agree – apply to a practice or policy that cut children off from their biological parents in a deliberate and calculated fashion. Speaking of human cloning, Ulla Wessels states that a clone might conceivably feel that her 'right to uniqueness' had been violated, and that this would be a serious consideration against the practice. Nevertheless, she explains, such a violation would take place only if the clone *wanted* to be genetically unique. I do not propose to analyse what seems to me to be a somewhat bizarre piece of reasoning, but wish instead to point out that in the same place Wessels notes another problem, one she seems to conflate with the first. She says that cloning might involve 'frustration of people's desire to know who they are and where they come from'. Moreover, simply hiding the relevant facts from the cloned offspring would, she thinks, be morally objectionable.[28]

It is this problem, quite distinct from the question of genetic uniqueness, that lies at the centre of the debate over reproductive cloning as well as other forms of artificial reproduction.[29] Consider the practices of adoption, or of sending children to orphanages, or fostering them out when their parents either do not want or are not capable of looking after them. Such practices do indeed tend to cut children off from their biological families. But natural law theory does not thereby judge them immoral, since they are regarded as harm-minimisation practices designed to alleviate the suffering caused to children without biological parents to look after them. It is never considered *good* to sever children from their biological roots; rather, it is an unwanted side-effect of giving children the support without which they will barely flourish at all. The loss of biological ties is not intended, it is not a means to anything, nor is it a goal of the practices just mentioned. Contrast such practices with one involving an Offspring Warehouse, where children were deliberately produced for the purpose of farming them out to any prospective parents who wanted them. Such a practice, which people still, one hopes, would find abhorrent (though its similarity to IVF is, to say the least, disturbing) requires as a necessary means that children be divorced from their biological parents and entire biological and genetic heritage, all of which contributes fundamentally to a child's sense of who she is and where she has come from,[30] as even such a friend of genetic engineering as Ulla Wessels seems to appreciate.

The NL theorist locates reproductive human cloning in the same ethical region as the Offspring Warehouse rather than that of the adoption agency. Instead of focusing exclusively, as consequentialists do, on whether a particular practice causes actual mental or physical harm here and now to a particular individual or set of individuals, the NL theorist's conception of harm embraces both this and a wider field of concern. Donor insemination and other forms of assisted reproduction involve secrecy and deception concerning a child's origins, prompting some offspring to launch legal action to gain access to records.[31] The NL theorist and CP bioethicist can agree, as does Wessels, that such deception is morally unacceptable. But the former goes further, holding that it is an injustice inherent in the practice whether or not identifiable individuals suffer specific, quantifiable harm from being deceived. Holding knowledge and family to be basic goods, the NL theorist claims that knowledge of one's familial origins is intrinsically good for a person whether or not lack of such knowledge distresses them. Such knowledge includes information about their ancestral, racial, national and geographical roots, their medical inheritance (including susceptibility to certain kinds of illness), and facts about who their siblings are and how many they have. The latter is important not simply because of the risk (serious enough though that be) that, absent such knowledge, an artificially produced individual might end up marrying one of them, but because every person has the right to information about who their biological family is now.

One can see that even if such secrecy were removed – and despite the best efforts of lobby groups this is unlikely, given the interests involved and the fear of what might happen if such information were routinely made available – the nature of the information currently held back is such that its wholesale release would cause immense social instability. Further, the more common assisted and artificial methods of reproduction become, the greater the potential instability. Now the CP bioethicist's response, for all her usual lip service to the problem of harm, is that what is really needed is a programme of education designed to inculcate the idea that all of the facts mentioned – concerning origins, family, medical inheritance, biological roots, and so on – are really not *so* important that concern for them should outweigh the advance of science. Mary Warnock, for example, accepts that when evaluating the ethics of reproductive cloning, one needs to ask: 'Would the possibility of human cloning be prejudicial to the good of society?'; and 'whether or not the child who was a clone would suffer'.[32] Yet the most she can come up with when answering the first question is that if cloning became widespread, there would be 'a damaging diminution of the human gene pool'. As for the second, she notes that society is becoming more tolerant of 'different kinds of families within which children may flourish', and hence that as long as the environment were loving, no suffering would result. There is no mention of secrecy and deception, and none concerning a child's right to know their biological and other origins, or whether lack of such knowledge intrinsically harmful.

Needless to say, there is no mention by Warnock or other supporters of cloning of other features of the practice not mentioned so far: its commodification of human life;[33] the multiple sacrifice of embryos in the trial-and-error process of perfecting the technique; the deformed and mutated human beings certain to be produced as a 'by-product' of the process; the unscrupulous and exploitative methods that would almost certainly have to be used to secure the continuing supply of that most crucial 'commodity', human eggs;[34] the possibility, if not likelihood, of there being biological fathers of thousands of children who do not know this (note that it is not merely the 'diminution in the gene pool' that makes such a phenomenon repugnant);[35] the export trade in gametes;[36] the continuing stories, only likely to be multiplied, of 'mix-ups' involving children of one race born to mothers or fathers of another race, embryos being 'lost', 'sold' to the wrong couple, and so on.[37] And all this without even beginning to mention the sorts of thing about which CP bioethicists are most often exercised, namely the actual, physical harm and unknown risks of harm caused to children produced by techniques, such as IVF, that fall short of cloning, thus indicating the exponentially greater risk of the latter.[38]

The reason CP bioethicists generally have little to say about the sorts of harm just mentioned (apart from the risk of actual physical harm noted at the end of the last paragraph) is that their approach to bioethics has no conceptual space for it. The undermining of basic goods, and of human goods in general, does not figure in their analysis, so they cannot come to grips with the idea that a practice may be prejudicial to society, and to particular individuals, by virtue of its tendency to undermine the goods that make possible the proper functioning of society and of the individuals within it. Even if secrecy and deception do not cause quantifiable harm to a particular person, that person can be wronged nevertheless, since human life does not flourish when knowledge of one's origins is suppressed. Further, secrecy and deception are built into the very conduct of certain biotechnological practices, and this of itself undermines trust and honest dealing in society, particularly between scientists and the public.

More generally, natural law theory holds to be wrong those practices which, of their essence, undermine basic goods. By saying 'of their essence' I mean that the practice itself is defined at least in part by its good-undermining features. For instance, the production of human beings in a way that, by its very nature, deprives them of their biological parents, is inherently wrong. But a practice might be such that, even if its good-undermining feature were not essential to it, that feature was so bound up with the practice as actually carried out that the practice *in its current form* would be inherently wrong. Such would be the case with, say, experimental techniques that undermined human life and health by being excessively risky given the state of current knowledge. The NL theorist generally evaluates such cases by use of the Principle of Double Effect, a kind of reasoning derided by CP bioethicists but eminently reasonable and defensible.[39] A practice might also be such that its practitioner had to *intend* to bring about its good-undermining feature in order to be able to carry out the practice: this would be true whether the feature was a specific goal of the practitioner, or whether bringing about the feature was a necessary means to realising some other goal, the operative principle being that he who intends the end intends the means. A scientist who, for instance, deliberately cloned thousands of offspring using his own somatic cells, say on the ground that the world needed more geniuses like him,[40] would exemplify the first case (a specific goal), as would genetic screening of babies for the purpose of aborting handicapped ones. The second kind of case (a necessary means) would be exemplified by the necessary 'overproduction' of embryos during IVF treatment, these embryos later being experimented upon or killed, or both.

When it comes to genetic engineering, then, the natural law approach is the same as it is for all other practices. The NL theorist wants to know what goods, if any, are undermined by it, whether the goods involved are personal or social (or both), how they are undermined, and what the intent is of the practitioners concerned. He does not discard or condemn a practice simply for being 'alien', or new, or experimental, or artificial. He might regard it as an acceptable 'second best' to some other practice, such as adoption instead of child-raising

by biological parents. Or he might regard it as an unacceptable alternative because of its very nature. Singer and Wells do not seem to understand this when they complain that the natural law objections to IVF
as unnatural should not preclude the NL theorist from regarding IVF as
a '"second best" method of conception [that] may still be far better
than none at all'.[41] The NL theorist has no qualms about using inferior
alternative methods, artificial or not, for achieving human goods.
What he objects to are those methods that themselves undermine
human goods in the very pursuit of them. Such is the case with IVF, so
most NL theorists believe, because it divorces sex from reproduction in
a way that undermines both goods, making the one a mere object of
pleasure without responsibility, and of the other a quasi-industrial
practice shrouded in deception, corrosive of family life and destructive
of untold numbers of commoditised human beings, some killed, others
experimental playthings, yet others in indefinite 'cold storage'. Contrary to the critics' charges, there is in this analysis no obsession with
sex and no unhealthy desire on the part of some ethicists to prescribe
how other people should behave in their private lives. The NL analysis
of sexuality is simply a part of the overall NL approach to human
goods in general. And the concern with certain intimate or private
practices is not with their *intimacy* or *privacy*, but rather with their
impact on human relations in general, in other words on the *common
good*, a concept that simply does not figure in CP bioethics except
when it is used to justify the onward march of scientific progress.

Application of natural law theory to specific questions

Stephen Clark, in a discussion of biotechnology and the concept of
natural integrity, poses the question of why a consequentialist should
not advocate the genetic engineering of 'deaf, blind, legless, microcephalic lumps' on which to experiment.[42] After all, this could minimise the amount of suffering experimental subjects had to experience
for the good of scientific knowledge. Yet our common intuition (one
hopes, once more) is that there would be something deeply repugnant
in such a practice. Clark locates the unease in the idea that it is permissible for humans significantly to reduce the capacities of independent living beings with their own *telē*, their own proper modes of
functioning, and to turn them into artefacts wholly dependent on
their makers. Especially insidious, he thinks, is the thought that such
'transgenic brutes' would owe their very existence to their artificers,
and so lack any grounds for complaint. (Since they themselves, by

their very characteristics, could not complain, the idea is supposedly that no human would have the right to complain on their behalf.) Elsewhere I have called this the problem of 'existential debt'.[43] Clark goes so far as to call such a practice an example of 'totalitarianism' towards the natural world.

The natural law theorist's general approach to Clark's imagined scenario is quite similar, but the emphasis is different. The former does not see the natural world as having the kind of independent being which prevents its being used for the benefit of man in quite significant ways.[44] NLT recognises human *dominion* over the natural world, though not the idea that humans may exploit it in any way they see fit: human exploitation of the natural world must conform to human nature's own requirements. Yet what are those requirements? Precisely the basic goods through pursuit of which human beings flourish. The wanton destruction of areas of natural beauty, for example, might undermine human pursuit of the basic good of aesthetic experience. Needless to say, one must analyse particular cases in order to decide on the legitimacy of a specific act, taking into account, for instance, the need for food, shelter, water, and so on. In this, as in all other matters, goods must be balanced and none may be directly undermined for the sake of any other. Humans dam rivers, clear forests, dynamite mountains, empty lakes – in this they despoil the landscape and override the 'independent being' of countless living things. But the NL theorist cannot lay down a blanket prohibition on such activities. Instead, he evaluates them according to the same set of principles that apply to any human action, considering, among other things, whether: the goods aimed at are proportionate to the harms risked (as in the case of pollution); what is at issue is the common good or private good – the former allowing more extreme interventions in the environment than the latter, but the latter potentially involving the good of private property; and whether a given intervention respects other human goods such as health and family life.

There is, however, plenty of room within NLT for a radical critique of certain kinds of intervention in the natural world based on an analysis of the style of human living that seems to make those interventions necessary. For instance, although NLT does not – or so I would argue[45] – recognise the existence of animal rights, it may well be that certain practices, such as the mass hunting of animals for the production of luxury goods, reflect an unhealthy materialism and vanity whose replacement by a more sober and rightly prioritised lifestyle would lead to the saving of countless animal lives. The same could be said with even more justification for the use of animals in the cosmetics industry or in the production of drugs aimed at alleviating the harm caused by self-inflicted, unnatural ways of living. (Think of obesity, adult diabetes and various auto-immune diseases whose rate of increase may well be due to the unnecessarily toxic environment we have produced and to the nutritionally deficient products masquerading as food that have become a staple for so many people.)

Supposing a cure for cancer could be generated from experiments on microcephalic animal lumps, then for all our understandable disquiet, NLT will not lay down a universal prohibition. But the NL bioethicist will immediately want to know much about the probability of success (as opposed to the probability of receiving a scientific IOU redeemable at some unspecified future time), the availability of other means, and how it is that humanity may have got itself into the position of appearing to need such a practice. Sometimes humanity has done nothing to have such a need: think of the millions of monkeys that have been killed for the sake of producing polio vaccine. In other instances, though, the supposed need for radical intervention in nature can be laid directly at the door of human avarice, vanity and other vices resulting from a disordered way of living.

Similarly, NL theory does not lay down a blanket prohibition on the use of genetic engineering for the improvement of food quality. For one thing, the NL theorist does not – and should not – claim that there is something intrinsically wrong with interfering in the natural genome of an organism in order to alter its characteristics for certain purposes or to produce strains that are more conducive to human exploitation. This has been done since the dawn of civilisation by the practices of domestication, selective breeding and hybridisation. Nevertheless, it must be asked what sort of situation has arisen that the very idea of, say, transgenic crops or animals should even become thinkable as a proposed solution to the problems of feeding the starving or of improving food supply to the affluent. Application of the Principle of Double Effect, for instance, requires that we consider whether, on the assumption that the good effect is proportionate, alternative practices are available that do not require the production of a given ill-effect. Mass transportation is a good effect whose benefits to humanity's quality of life, at least in the current state of things, is proportionate to the harm caused by pollution. But this does not absolve humanity of finding ways to minimise pollution while still securing the benefits of transportation.

Similarly, given the pressing problem of mass hunger, policy-makers in general, and farmers and food manufacturers in particular, are not absolved from the requirement to find ways of solving that problem without resort to risky technological practices that could lead to serious harm. The same applies to improving the quality of food and efficiency of food production for the affluent. Hence it is plausible to argue that research on methods of organic food production is a far more pressing need than seeking to use technology in a way that simply reinforces practices involving the production of poor quality, nutritionally deficient, environmentally contaminated food. The NL theorist should try to locate a practice within a broader picture of man's relationship to nature: for instance, how human health is best promoted by the avoidance of deficiency in diet and toxicity in the environment. This is why the NL theorist is likely to baulk at the idea of transgenic crops produced to enable the use of ever more and stronger herbicides and pesticides. The issue is not whether the very idea of a transgenic food source should be anathema because it 'interferes with nature', but whether something has gone wrong with human living that makes resort to such risky practices thinkable in the first place. It is arguable that in the case of ancient practices such as cross-breeding and domestication, mankind has generally worked with a body of knowledge, accumulated over thousands of years of experience, of the ebb and flow of the seasons, changes of climate, animal behaviour and the like, and has possessed genuine wisdom concerning man's interaction with the delicate ecological balance. This valuable corpus of experience, handed down from generation to generation, of people who worked on and with the land has enabled a gradual development of agricultural practice that generally respected, indeed revered, the power of nature to shape man's life.

Yet even here mankind has reaped the whirlwind of thoughtless and exploitative change that has led to desertification, pollution and monocultures destructive of the very natural variety that conduces to human health and wellbeing. How much greater is the risk in respect of technologies that are often *designed* precisely to work *against* nature, for instance by subverting natural disease resistance in favour of an engineered resistance that enables yet more toxic products to be introduced into the environment with the consequences for human health that are only too apparent? Perhaps, in thousands of years – assuming our species survives that long – genetic engineers will have accumulated an equivalent body of knowledge enabling them to build on, rather than replace, the venerable traditions that have for the most part served

humanity well. In the meantime, however, we cannot shirk the responsibility of asking, for any biotechnological innovation, how exactly it is
supposed to serve mankind and the environment on which we depend,
in the light of a complete picture of what fulfils us. In short, NL theory
demands a radical critique of economic and social policy, one that that
looks for solutions to problems that are deep-rooted in human nature
and demand more than a mere technological fix.

The same considerations apply to such practices as adult stem cell
research and somatic cell therapy. In both cases, the professed aim of
biotechnologists is to improve the human condition. In so far as this
means the elimination of disease, the NL theorist has no criticism to
make, but insists only on a full evaluation of the proportionality
between the good aimed at and the harm risked. The evaluation is
decidedly *not* consequentialist in character.[46] NL theory insists, for
instance, that in any evaluation of proportionality all rights must be
respected, i.e. none must be violated. An obvious example of violation
would be the use of adult stem cells in research without the consent of
their possessor. Another would be the production of human life for the
purpose of experimentation designed to develop somatic cell therapies,
whereby the humans concerned were treated as 'products' useful for
exploitation.

In the case of so-called 'designer babies', NL theory insists again on a
judgement concerning the *practice* in the light of what it will probably
do to human relations. In the sorts of case that have arisen, parents
have invariably said that they wanted another child anyway, and
regarded the fact that the new child was a genetic match for a sick
older sibling as an additional benefit. But the very idea that a child
could be produced for spare parts, even if this was only an 'additional'
purpose, is corrosive of parent-child relations. It is not a mere 'slippery
slope' argument lurking behind such a reaction, as though the real
objection were that perhaps, sometime in the future, parents will start
producing children *solely* for spare parts. Rather, the problem lies in the
very practice from the outset, in the idea that children can be manu-
factured to order, for exploitation of their body parts. The child is no
longer seen wholly as a gift, but partly as a product.

This aspect of the NL critique does not, of course, refer to the complete ethical context in which 'designer babies' are produced. Perhaps
the day will come when couples will know in advance when is the
most opportune time to conceive a child with a view to producing one
with the right genetic match for a sick sibling. For the moment,
however, the practice requires the screening of numbers of embryos for

the correct match and the selective destruction of those that do not fit the bill. This is why inverted commas were placed around the word 'additional' earlier: for the very use of screening tends to belie the notion that finding a genetic match is not the primary – or at least equally important – purpose of producing the child in most cases. Perhaps this will not be so in every situation, but one can ask, for instance, whether a couple who found that genetic screening for their offspring suddenly became unavailable would go ahead with the pregnancy nevertheless. If not, one would at least have evidence that finding a match was a primary purpose. If they would still go ahead, although this would be evidence that finding a match was *not* a primary purpose, it would still be consistent with there being a goal of finding a match even if that goal was of only secondary importance. One must, nevertheless, take into account that counterfactual tests of intention, though bearing prima facie evidential weight in most cases, are still defeasible.[47] Evidence of intent aside, for the NL theorist no biotechnological practice is permissible if rights are not respected, and so the fact that lives must be destroyed for genetic matches to be found is reason enough for prohibiting such a technique.

The same can be said for somatic cell engineering, germ line engineering and all kinds of non-therapeutic genetic enhancement or elimination of undesired characteristics. The NL theorist will wonder what sort of attitude to human life could lead to the thought that one might, say, allow the genetic engineering of 'wonderwomen and supermen', to use the title of a book by John Harris. Speaking of sex selection – probably the mildest of the sorts of eugenicism countenanced by Harris and others – he comments: 'Children of either gender might well resent being "made to order" and might resent their parents for so acting. However, I am far from convinced that there are any cogent arguments for the entitlement to be protected from feeling resentment and the children in question might find it salutary to reflect that but for their parents' choice, they themselves would not exist at all.'[48] The typical CP bioethical reaction to such practices is to frame the issue in terms of contingent feelings of hurt, neglect or resentment. Even if the resentment were widespread, the CP response is simply that such children would need more education, that social attitudes would have to change. Hence the CP bioethical position can be seen as ad hoc, subject-changing and apparently immunised against criticism.

The NL theorist, on the other hand, resists the invitation to conduct the debate in term of what this or that child might happen to feel. NL theory asks instead what sort of attitude human beings should have to reproduction in general, and how society should view its members. To have a child for what it can do, or be, rather than as an expression of mutual love between a couple, is itself a distortion of human relations. In the case of sex selection in China, for instance, it is no accident that a preference for males goes hand in hand with widespread female infanticide.[49] Nor is it an accident that in countries where male off-spring are prized over females, women are often denigrated and humil-iated for not 'doing their duty'. The CP bioethicist will, of course, reply that such attitudes to women who do not produce the desired male are due to biological ignorance, which putting sex selection into the hands of a scientist will neutralise.

Let us leave aside the fact that using technology to select sex will do nothing to prevent infanticide when science fails to produce the desired result. Leave aside also the fact that there will always be some-one to 'blame' if the baby with the desired gender does not materialise (usually everyone but the father). What the natural law theorist will want to know is how engineering the sex of one's offspring, let alone any other 'desirable' characteristics such as intelligence or potential athleticism, is supposed to be consistent with the attitude of *uncondi-tional* love and affection essentially involved in the promotion and respect of the basic good of human life? By definition the imposition of conditions upon the features of one's potential offspring is *inconsist-ent* with such an attitude. Of course, it is the case that in particular instances a couple who have selected the sex of their offspring, or who have chosen a child with the physical potential to be an Olympic gold medallist, might well show that child love and affection whatever the actual outcome. But this no more supports a general principle in favour of eugenicism than the fact that incest or paedophilia some-times causes no obvious harm to anyone supports the legitimisation of either of these practices.

NL theory is concerned not just with how things are in respect of particular individuals but with the *common good*, in particular with those sorts of human relation and interaction that are conducive to a harmonious society built on charity, justice and mutual respect. Specifically, a society in which respect for and protection of its most vulnerable members is not a priority, is a society that NL theory regards as perverted. Not only does the imposition of conditions on reproduction tend to corrode such virtues, but they are positively undermined by the practices that make sex selection and modern eugenics possible in the first place: the manufacture of human beings by technicians; embryo experimentation, exploitation for spare parts, and destruction; the elimination of young human beings who are 'unfit' or undesired for a given purpose; the commericalisation of gametes; and the severing of human beings from their biological and ancestral inheritance. Needless to say, the Nazi experience proves that such technology need not be available for a eugenic agenda to be carried out – it is the underlying attitudes that fertilise technological development. Although the motives may have changed, and the science may be more rigorous, the same attitude – the fundamental conditionalisation of human life and reproduction – animates contemporary thinking in the vast area of bioethics devoted to promoting the genetic engineering of human beings.

Conclusion

The above discussion is no more than a gesture in the direction of a natural law critique of contemporary genetic technologies. Since natural law theorists are yet to provide the sort of complete analysis that is required, there are many questions remaining unanswered. One of the crucial problems is what is sometimes called the problem of 'dirty hands'. For genetic engineering involves some technologies that may prove beneficial to human life without violating rights or goods – adult stem cell research and somatic cell therapy for the cure of inherited and other disease being two of the most obvious. Yet even the most intrinsically ethically acceptable practices may only have been made possible on the back of other practices that are evil (consider beneficial technological by-products of the torture of innocents or experiments on unwilling people). Still other technologies, although in themselves unobjectionable, may be of a kind that can only take place within a research infrastructure that condones illegitimate practices such as, for instance, embryo experimentation or abortion. Biotechnology is not merely a collection of research efforts going on in various parts of the world – it is a vast industry supported by billions of pounds of government and private funding, with an intricate network of institutions and regulation.

Can such a network be, as it were, purified? Is it technically possible, and is society ready, to undertake the sort of critique that turns the infamous 'yuck factor' into a set of concrete proposals for disengaging legitimate from illegitimate research – for example, that into adult stem cells from that into embryonic stem cells? And is it feasible to turn these proposals into legislative restrictions that prohibit impermissible research and direct public and private funding into acceptable channels? With contemporary bioethics in the state that it is, and with the consequentialist-pragmatist approach dominating both academic and popular debate as well as the formulation of public policy, there is no obvious cause for optimism. What is clear, in my view, is that if natural law theorists – whose ideas are attracting more interest at the normative level – provide the sort of *applied* ethical critique needed for a reappraisal of the direction in which biotechnology is going, at least one condition necessary for a reconfiguration of social attitudes will have been fulfilled.[50]

Notes

1 For a collection of papers defending this view of consequentialism, see D. S. Oderberg and J. A. Laing (eds.) (1997) *Human Lives: Critical Essays on Consequentialist Bioethics* (Basingstoke: Macmillan).

2 P. Singer and Deane Wells (1984) *The Reproduction Revolution* (Oxford: Oxford University Press), pp. 36–41.

3 Singer and Wells, *The Reproduction Revolution*, p. 37.

4 Singer and Wells, *The Reproduction Revolution*, p. 37.

5 Singer and Wells, *The Reproduction Revolution*, p. 37.

6 It might be objected that by this way of talking I am helping myself to a teleological view of nature that is rejected *ab initio* by most philosophers. It is true that this view of nature is largely a presupposition of the argument that follows, there being no space for a direct justification of it here in terms, for instance, of the remarkable adaptation of organisms in nature. But it is also my intention to show that what follows from such a view gives it indirect support. In particular, if the sort of elaboration of teleology that natural law theorists give is at least consistent with certain strong moral intuitions, that is an indirect argument for such an elaboration. Moreover, as we will see, teleology is difficult to escape on both sides of the bioethical debate, however one might elaborate one's teleological view.

7 Singer and Wells, *The Reproduction Revolution*, p. 37.

8 Singer and Wells, *The Reproduction Revolution*, p. 37.

9 Singer and Wells, *The Reproduction Revolution*, pp. 38–9.

10 On this topic, see D. S. Oderberg (2000) *Moral Theory: A Non-Consequentialist Approach* (Oxford: Blackwell), pp. 3–9.

11 Singer and Wells, *The Reproduction Revolution*, p. 39.

12 Singer and Wells, *The Reproduction Revolution*, p. 41.

13 For a recent critique, see C. J. Martin (2004) 'The Fact/Value Distinction', in D. S. Oderberg and T. Chappell (eds.), *Human Values: New Essays on Ethics and Natural Law* (Basingstoke: Palgrave Macmillan), pp. 52–69. See also Oderberg, *Moral Theory*, pp. 9–15.

14 M. J. Reiss and R. Straughan (1996) *Improving Nature? The Science and Ethics of Genetic Engineering* (Cambridge: Cambridge University Press), p. 63.

15 J. Cottingham (2004), '"Our Natural Guide ...": Conscience, "Nature", and Moral Experience', in Oderberg and Chappell (eds.), *Human Values*, pp. 11–31.

16   I will now use terms such as 'victim' and 'suffer' with the proviso that this not be understood as injecting a circularity into my discussion of morality and harm. One can undertake the discussion without using such words, but their use is convenient.

17   For more on this, see S. Uniacke (2004) 'Harming and Wronging: The Importance of Normative Context', in Oderberg and Laing (eds.), *Human Lives*, pp. 166–83.

18   For more on this problem, see Jacqueline A. Laing (2004) 'Innocence and Consequentialism: Inconsistency, Equivocation and Contradiction in the Philosophy of Peter Singer', in Oderberg and Chappell (eds.), *Human Values*, pp. 196–224, at pp. 209–12.

19   M. Häyry (1994) 'Categorical Objections to Genetic Engineering – A Critique', in A. Dyson and J. Harris (eds.), *Ethics and Biotechnology* (London: Routledge), pp. 202–15, at p. 202.

20   Häyry, 'Categorical Objections to Genetic Engineering', p. 202.

21   Häyry, 'Categorical Objections to Genetic Engineering', pp. 209ff.

22   U. Wessels (1994) 'Genetic Engineering and Ethics in Germany', in Dyson and Harris (eds.), *Ethics and Biotechnology*, pp. 230–58.

23   Wessels 'Genetic Engineering and Ethics in Germany', p. 240.

24   Wessels 'Genetic Engineering and Ethics in Germany', p. 245.

25   Reiss and Straughan, *Improving Nature?*, p. 67.

26   See, for instance, J. Finnis (1980) *Natural Law and Natural Rights* (Oxford: Clarendon Press); T. Chappell (1998) *Understanding Human Goods* (Edinburgh: Edinburgh University Press); M. Murphy (2001) *Natural Law and Practical Rationality* (Cambridge: Cambridge University Press); Alfonso-Gómez-Lobo (2002) *Morality and the Human Goods* (Washington, DC: Georgetown University Press); see also Oderberg, *Moral Theory*, ch. 2, and (2004) 'The Structure and Content of the Good', in Oderberg and Chappell (eds.), *Human Values*, pp. 127–65.

27   Hence the disingenuousness of the remark by Singer and Wells, that '[t]he difficult question of course is how the laws of nature are to be identified, and different natural law theorists have given different answers', *The Reproduction Revolution*, p. 39.

28   Wessels 'Genetic Engineering and Ethics in Germany', p. 243.

29   For more on this, see J. A. Laing and D. S. Oderberg (forthcoming) 'Artificial Reproduction, the "Welfare Principle", and the Common Good', *Medical Law Review*.

30   Cf. Draper's chapter in this volume.

31   See, for example, Joanna Rose (1999) 'The Response of an Adult Donor Insemination Offspring to the Article "The Psychology of Assisted Reproduction – or Psychology Assisting its Reproduction?"', *Australian Psychologist*, 34, p. 220; Christine Whipp (1998) 'The Legacy of Deceit: a Donor Offspring's Perspective on Secrecy in Donor Assisted Conception', in Eric Blyth, Marilyn Crawshaw and Jennifer Speirs (eds.), *Truth and the Child 10 Years On: Information Exchange in Donor Assisted Conception* (Birmingham: British Association of Social Workers). For further information, see the notes in Laing and Oderberg, 'Artificial Reproduction'.

32   M. Warnock (2003) 'What is Natural? And Should We Care?', *Philosophy*, 78, pp. 445–59, at pp. 457–8. Quotations are from these pages.

33 Cf. Stephen Wilkinson, chapter 2 in this volume.

34 It currently takes hundreds of eggs to produce one cloned mammal, but hyper-ovulation of consenting women directly involved in production of the offspring will not produce a large enough supply (the unpleasantness of the procedure being enough of a deterrent in itself). Hence the biotech industry will have to resort to paying poor women, especially in less developed parts of the world where they will be in greatest need and least likely to understand what is being done to them.

35 The following story, from London's *Metro* newspaper (27 October 2003), is likely to be followed by many similar ones in the future: 'Fears have been expressed that a psychopath who murdered his baby daughters could have fathered hundreds of children through sperm donations, *Metro* reports. Heine Nielsen, who is now serving life imprisonment, is thought to have made 520 donations to a Danish sperm bank which supplies 40 countries. One expert urged women who believed themselves to be pregnant with his children to have abortions in case they had Nielsen's genes.'

36 For example, Ole Schou is chief of Danish sperm bank, Cryos International Sperm Bank Ltd: 'Cyros dominates the Scandinavian market from its headquarters in Aarhus, Denmark. In the early 1990s, the country began looking abroad for ways to expand its business and now exports to 25 countries, including Australia, Eastern Europe, and the US. It market three grades of sperm, including "Extra" grade, which contains twice as many sperm as the standard grade and exhibits the highest levels of motility, a measure of sperm's ability to reach its target' (Pascal Zachary, *Wall Street Journal*, 6 January 2000).

37 The embryologist Sammy Lee has recently written that mix-ups are a regular feature of clinical practice, that commissioning parties are sometimes kept in ignorance of this fact and further, that mix-ups are occasionally deliberate, for instance so that a 'deserving' couple be provided with a child: BBC News Online, 24 July 2002; *Sunday Telegraph*, 10 November 2002. For a case involving the birth of black children to white parents, see *The L Teaching Hospitals NHS Trust* v. *Mr A, Mrs A, YA, ZA (By their Litigation Friend, The Official Solicitor), The Human Fertilisation and Embryology Authority, Mr B, Mrs B*: High Court of Justice, Queen's Bench Division, 26 February 2003, [2003] 1 FLR 1091, [2003] EWHC 259 (QB).

38 For emerging evidence of the harm caused by IVF, see M. Hansen, J. J. Kurinczuk et al. (2002) 'The Risk of Major Birth Defects after Intracytoplasmic Sperm Injection and In Vitro Fertilization', *New England Journal of Medicine*, 346, pp. 725–30. This study found that IVF offspring have twice as high a risk of a major (e.g. chromosomal or musculoskeletal) birth defect as naturally conceived infants.

39 See Oderberg, *Moral Theory*, ch. 3; T. Chappell (2002) 'Two Distinctions that Do Make a Difference: The Action/Omission Distinction and the Principle of Double Effect', *Philosophy*, 77, pp. 211–34; M. Murphy (2004) 'Intention, Foresight, and Success', in Oderberg and Chappell (eds.), *Human Values*, pp. 252–68.

40 Thoughts of sperm banks for Nobel Prize winners spring to mind, such as the Repository for Germinal Choice in California.

41 Singer and Wells, *The Reproduction Revolution*, p. 38.
42 S. R. L. Clark (2004) 'Natural Integrity and Biotechnology', in Oderberg and Chappell (eds.), *Human Values*, pp. 58–76, at p. 67.
43 Laing and Oderberg, 'Artificial Reproduction'.
44 There are dissenters from this view: see Chappell, *Understanding Human Goods*, p. 40, where he places the natural environment among the basic goods, but without argument.
45 See further D. S. Oderberg (2000) *Applied Ethics* (Oxford: Blackwell), ch. 3.
46 For more on this, see Oderberg, *Applied Ethics*, pp. 97–101.
47 See further Oderberg, *Applied Ethics*, pp. 118–21.
48 J. Harris (1992) *Wonderwoman and Superman: The Ethics of Human Biotechnology* (Oxford: Oxford University Press), p. 160.
49 In September 1997, the World Health Organisation's Regional Committee for the Western Pacific issued a report claiming that 'more than 50 million women were estimated to be "missing" in China because of the institutionalized killing and neglect of girls due to Beijing's population control program that limits parents to one child' (see Joseph Farah (1997) 'Cover-up of China's Gendercide', *Western Journalism Center/Free Republic*, 29 September); http://www.freerepublican.com/forum/98896.htm. Accessed 23 May 2005. A similar story is true for India.
50 I am grateful to Nafsika Athanassoulis for helpful comments on a draft of this chapter.

6
Autonomy, Inducements and Organ Sales

James Stacey Taylor

Introduction

It is widely known that the number of organs that become available for transplant each year falls far short of the number that are needed. Various methods of alleviating this shortage have been proposed, ranging from increasing public awareness of the organ shortage to encourage donation to introducing a policy of presumed consent, in which organs can be harvested from persons after their death unless they have expressly forbidden this. With one notable exception, all of the proposed methods of alleviating the chronic shortage of available transplant organs enjoy widespread support, even if they might also suffer from similarly widespread opposition. This exception is the proposal that markets should be used to procure additional transplant organs. This proposal has, as Janet Radcliffe Richards has noted, been condemned by almost all who are involved in the discussion of how to increase the number of available transplant organs, irrespective of their political or ethical commitments.[1] The arch-conservative Leon Kass, appointed by George W. Bush to be the Chairman of the President's Council on Bioethics, for example, condemns the use of markets to secure an additional supply of transplant organs with as much venom as its more politically radical opponents, such as Nancy Scheper-Hughes and Lawrence Cohen.[2] And if the debate over organ sales makes for strange political bedfellows, it makes equally odd bedfellows of professional ethicists. Kantians and consequentialists alike have argued that using markets to procure human transplant organs is morally impermissible.[3]

As well as there being widespread agreement that the use of markets to procure human transplant organs is morally impermissible there is also widespread agreement as to why this is so. Many opponents claim that such markets would serve to compromise the autonomy of those who would participate in them as vendors. Since respect for personal autonomy is the pre-eminent value in contemporary bioethics, the use of markets to procure human transplant organs is morally impermissible.[4] In this chapter I will argue that the widespread view that respect for autonomy militates against the view that using markets to procure human transplant organs is mistaken. In particular, I will argue that the widespread claims that a current market in human transplant organs would compromise the autonomy of potential vendors through enabling their poverty to coerce them into selling is false, as is the claim that such a market would compromise the potential vendors' autonomy by providing them with irresistible financial inducements for their organs.

An autonomy-based pro-market argument – and some initial caveats

At first sight the claim that respect for autonomy militates *against* instituting a market to secure human transplant organs might seem odd. After all, the institution of such a market would provide persons with an additional option – the option to sell one of their organs. As such, this would provide persons with an additional way in which they might choose to exercise their autonomy. Since this is so, it appears that the principle of respect for autonomy should *support* the view that the use of markets to secure human transplant organs is morally permissible, rather than militate against it.[5]

The intuitive appeal of this autonomy-based pro-market argument, however, rests on two implicit premises. The first is that the vendor would be autonomous with respect to the sale of his organ. The second is that the option to sell an organ would not be an autonomy-compromising constraining option; that is, that the option to sell an organ would not be an option that, if chosen, would be likely to lead to the person who chose it suffering from compromised autonomy as a result. I have defended the second of these implicit premises elsewhere, and so will not address the question of whether the option to sell an organ is a constraining option here.[6] The first of these premises, however, has yet to be fully defended.[7] I will undertake this defence here.

Before moving to my arguments, several caveats must be put in place. First, I will focus here on the question of whether markets for human transplant kidneys are morally permissible, rather than on the broader question of the morality of markets in human body parts *per se*. Since kidneys are both paired and non-renewable they fall somewhere on the moral scale between replenishable body parts (such as blood, plasma and semen) on the one hand, and single irreplaceable body parts such as hearts or whole livers on the other.[8] Second, I will defend the moral permissibility of a *current* market for human kidneys, in which the vendor's kidney is removed from his body while he is still alive, rather than a cadaveric market (in which the legal possessors of a corpse sell its organs) or a futures market (in which persons sell their organs for possible removal after they die). I focus on defending a current market for human kidneys because such a market is held to be the most morally objectionable from the point of view of one who is concerned with protecting the autonomy of those who would be moved to sell their kidneys. As such, then, if I can show that a current market in human kidneys is morally permissible from the point of view of a defender of autonomy I will also have shown that there is good reason to believe that the other two types of markets in kidneys are permissible also. Finally, in this chapter I will be concerned only to defend the view that it is morally permissible to use markets to *procure* human transplant kidneys. I will not address the separate issue of whether a market should also be used to *distribute* the kidneys thus procured.[9]

Autonomy and identification

Prior to arguing that respect for autonomy requires that persons be allowed to sell their kidneys for transplantation I must clarify how the concept of autonomy is to be understood in the context of this argument. This clarification is important, for despite the attention that autonomy has received in recent years from both autonomy theorists and applied ethicists there is still a great deal of confusion surrounding this concept. Much of this confusion has been generated by a widespread failure to recognise that the concept of autonomy is *not* coextensive with the concept of identification.

The concept of identification, of what it is for a person to identify with her effective first-order desires, was first developed by Harry Frankfurt is his seminal paper 'Freedom of the Will and the Concept of a Person', in which he was primarily concerned to provide an 'analysis of the concept of a person'.[10] Frankfurt argued that a characteristic peculiar of persons is that they 'have the capacity for reflective self-evaluation'.[11] In particular, Frankfurt argued that for an agent to be a person it is essential that he have second-order volitions. That is, it is essential that (at least sometimes) he reflect on his effective first-order desires to determine whether he both wants to have them, and wants them to be effective in moving him to act (or omit to act). With this account of personhood in hand, Frankfurt noted, 'There is a very close relationship between the capacity for forming second-order volitions and another capacity that is essential to persons' – the capacity to enjoy or lack freedom of the will.[12] Frankfurt then argued that 'the statement that a person enjoys freedom of the will means ... that he is free to will what he wants to will, or to have the will he wants'.[13] Since Frankfurt understands the concept of the will in Hobbesian terms (i.e. a person's will is that appetite of aversion that leads to action or omission)[14] a person possesses freedom of the will if he is able to secure 'the conformity of his will to his second-order volitions'.[15] For Frankfurt, then, a person will enjoy freedom of the will when he *identifies himself* with his effective first-order desires; that is, when he ensures that his effective first-order desires are the desires that he both wants to have and wants to be moved by.

Frankfurt's initial account of what it is for a person to identify with her effective first-order desires has been heavily criticised.[16] This has led him both to modify it and to clarify it.[17] The point of outlining Frankfurt's initial account of identification here, however, is neither to praise it nor to bury it. Rather, it is to show how the concept of identification is distinct from the concept of autonomy, so as to clarify how autonomy is to be properly understood. This clarification is especially important since Frankfurt's account of what it is for a person *to identify with* her effective first-order desires has been widely understood to be an account of what it is for a person *to be autonomous with respect to* her effective first-order desires.[18] From the above exegesis of Frankfurt's views it is clear that the question of whether a person *identifies with* his effective first-order desires concerns both his status as a person, and the issue of whether he is moved to act by first-order desires that he desires to act upon. Thus, if it is possible for a person to lack autonomy and yet *still* be moved to act by a desire that he volitionally endorses, then the concepts of identification and autonomy are not coextensive. And this is indeed possible. For example, a person who is deceived by another into desiring to perform a certain action will not be self-directed, and so will not be *autonomous*, with respect to either his desire to perform the action in question, or the action itself. It will not be he, but his deceiver, who is directing him to have the desire and to perform the action. However, that a person is deceived into performing an action that he would not have otherwise performed is no bar to his *identifying* with the effective first-order desire that moves him to act. For example, Othello certainly identified with his desire to smother Desdemona even though he was deceived into so acting by the machinations of Iago. Othello was acting as a person when he smothered Desdemona, and he smothered her of his own free will. Of course, one might claim that cases in which a person is deceived into endorsing an effective first-order desire serve as counterexamples to (what is widely taken to be) Frankfurt's account of identification-as-autonomy, rather than as showing that identification and autonomy are distinct concepts. There are two responses to this charge. First, it is clear that Frankfurt is developing an account of what conditions must be met for a desire to be attributable to a person as her own desire, such that if she acts on that desire, then she is acting of her own free will. In this research project Frankfurt makes no mention of the concept of autonomy.[19] As such, unless arguments are forthcoming that show that an account of what it is for a person to identify with a desire *must also* be an account of what it is for a person to be autonomous with respect to that desire there is no reason to believe that Frankfurt is attempting to provide an analysis of the concept of autonomy. To assume that he is, and thus to assume that examples of deception and manipulation can *ipso facto* be used as counterexamples to his account, is simply to beg the question. Second, unless there is good reason to assimilate the concept of free will to the concept of autonomy these concepts should be kept separate from each other. Not to do this could lead to difficulties in analysing these concepts, for if they addressed while commingled the intuitions that surround the use of one of these concepts might be drawn upon illegitimately (and unknowingly) while analysing the other. For example, as I noted above, if one holds Frankfurt's analysis of identification to be an analysis of autonomy one might be tempted to reject it on the grounds that a person could be manipulated into identifying with a certain desire. Yet unless arguments are forthcoming that show that these are *not* distinct concepts such a rejection would be premature.

What, then, can we learn from this discussion that will be useful in the context of the defence of the first implicit premise in the above autonomy-based argument in favour of using markets to procure human transplant organs? First, it is clear that this argument should eschew the approach, unfortunately so common in contemporary applied ethics, of drawing on an account of identification (usually

Frankfurt's) as though it were an account of autonomy, and then using it either to support one's position, or else to support one's view that autonomy is not morally relevant. (The first way of drawing on the concept of identification in this way is exemplified by the common claim 'The practice that I object to would/would not violate autonomy understood in this way...', while the second is exemplified in the claim 'Since patients don't [e.g.] endorse their effective first-order desires in this way and yet we still should respect their choices it is clear that autonomy is not important in medical ethics'.)[20] The arguments that I develop below, then, will not consider whether or not the typical potential kidney vendor would volitionally endorse his effective first-order desire to sell a kidney. Second, given that autonomy and identification are distinct concepts the fact that there is currently much debate over how to understand the concept of identification should not make one hesitate to use the concept of autonomy as the foundation for a moral argument, as one might were these two concepts coextensive. Finally, and most importantly, the above discussion of autonomy and identification has made it clearer how the concept of autonomy is to be understood. That a person (for example) endorses her effective first-order desires is not sufficient for her to be autonomous with respect to them, nor is it sufficient for her to be autonomous with respect to the actions that they move her to perform. Rather, to be autonomous with respect to her desires and actions it is necessary for a person to be *directing herself* to form, or perform, them, rather than to be (wittingly or unwittingly) directed to do so by another. A person is thus autonomous with respect to her desires and her actions if it is she, and not another agent, who controls which desires she forms and which actions she performs. Moreover, just as it is not sufficient for a person to identify with her effective first-order desires for her to be autonomous with respect to them, that she is autonomous with respect to her effective first-order desires is not sufficient for her to be autonomous with respect to the actions that such desires move her to perform. For example, it is possible for a person autonomously to desire not to resist another's usurpation of control over her actions. A person faced by a gunman might autonomously desire not to resist acting on his orders, for she might consider satisfying this desire to be her best option. However, in satisfying this desire it will not be she, but he, who is directing her actions. She will thus not be autonomous with respect to the act that she performs to satisfy the effective first-order desire that she is autonomous with respect to.

The argument from coercion

With this clarification of how autonomy is to be understood in place it is time to assess the arguments that are ranged against the view that respect for autonomy requires that market mechanisms be used to procure human transplant kidneys. The first and most prevalent of these anti-market arguments is the argument from economic coercion.[21] According to the proponents of this argument, markets in human transplant kidneys would serve to compromise the autonomy of many potential vendors by enabling their poverty to coerce them into selling their kidneys.[22] A person who has been coerced into performing an action suffers from a diminution in her autonomy with respect to it. So, if persons are coerced by their poverty into selling their kidneys then they will suffer from compromised autonomy with respect to their vending actions. Accordingly, since the typical vendor would be coerced into selling a kidney by his poverty and so would suffer from a diminution in his autonomy with respect to his vending action it seems that respect for the autonomy of the potential vendors should support the continued prohibition of markets for human kidneys.

This argument in favour of the continued prohibition of markets in human kidneys is clear, elegant and persuasive. Unfortunately, it is also fatally flawed. This is because its proponents overlook the reason why a person who is coerced into performing an action thereby suffers from a diminution in his autonomy with respect to it. From the above discussion of autonomy and identification it is clear that a person who is coerced into performing an action suffers from a diminution in his autonomy with respect to that action (although not with respect to the desire that moved him to perform that action) to the extent that he ceded *control* over his actions to the person who was coercing him. Control, however, is an intentionally characterised concept. A person can thus only cede control to an intentional agent. For a person to suffer from a diminution in her autonomy with respect to an action that she is coerced into performing, then, she must be coerced by an intentional agent, for she is conceptually unable to cede control to any other entity. Given this, it is conceptually impossible for a person to suffer from a diminution in her autonomy with respect to any action that she has been 'coerced' into performing by her poverty, for a person's economic situation is not something that she can cede control over her actions to. Thus, even if a person decides to sell a kidney only because he would otherwise starve, he does not thereby suffer from any diminution in his autonomy. It is not his poverty, but he, who is directing the performance of his actions.[23]

The arguments from irresistible offers

Outlining the arguments

The argument from coercion fails because its proponents were unable to identify an intentional agent to whom the potential kidney vendor cedes control of his actions in deciding to sell. This, however, is not a problem that besets a related cluster of autonomy-based objections to allowing current markets in human kidneys: those that focus on the claim that the offer to purchase a kidney would be, for many potential vendors, an irresistible offer. Although the arguments that are based on the claim that the price offered for a live human transplant kidney would prove to be an autonomy-compromising irresistible offer for many potential vendors are not neatly delineated in the literature on the morality of markets in human body parts, they can be divided into four main types. The first and simplest of these arguments is based on the claim that, for some persons, the price offered for the purchase of a kidney would literally be irresistible. For such persons allowing a current market for human transplant kidneys would lead them to sell their kidneys irrespective of whether they autonomously wished to sell or not. Thus, for those persons for whom the price offered for their kidneys proved to be irresistible and yet who did *not* autonomously wish to sell, a market for kidneys would lead to the diminution in their autonomy through enabling them non-autonomously to sell their kidneys.[24] The second of these arguments from irresistibility focuses on the possibility that a market for human kidneys would lead to some persons suffering from weakness of will. Such persons would not believe that the sale of one of their kidneys was the right action for them to perform, yet they would be so drawn by the inducements offered to them that they would see against their better judgement.[25] Since a person who acts out of weakness of will suffers from a diminution in her autonomy, instituting a current market for human kidneys would lead to some persons suffering from compromised autonomy who would not have otherwise done so.

The third argument that is based on the claim that some potential vendors will find the price offered for their kidneys irresistible is subtler than the first two. The proponents of this argument from irresistibility first note that, for some potential vendors, the amount of money that they could gain for their kidneys would be significantly higher than that which they could secure any other way. Since this is so, they continue, the option of selling their kidneys would, for such potential vendors, render ineligible the other options that such vendors would have otherwise pursued through drastically increasing the opportunity costs associated with their pursuit. Rather than having several eligible options open to them, then, after the introduction of a current market for human transplant kidneys such potential vendors would only have one: the option of selling their kidneys. The introduction of this option would thus diminish these potential vendors' ability to exercise their autonomy by reducing the number of eligible options that they possess. Thus, the proponents of this version of the argument from irresistible offers conclude, respect for the autonomy of the potential vendors in a current market for human kidneys militates against such a market. The fourth – and most interesting – of this family of arguments is based on the claim that some of the potential kidney vendors would be precluded from exercising their autonomy owing to their suffering from a form of motivational ambivalence generated by the tempting offers that they would receive for their kidneys.[26] The proponents of this argument first note that a person who is suffering from temptation is likely to experience motivational ambivalence. That is, he is likely to be as strongly motivated *not* to perform the act that he is tempted to do, as he is motivated to perform it. Such ambivalence would preclude a person who experiences it from exercising his autonomy, since it would preclude him from having a motivationally unified self to ground his *self*-direction. The proponents of this argument then note that many would-be kidney vendors *would* be tempted to sell, and so would experience such motivational ambivalence. Since this is so, they conclude, respect for autonomy requires that current markets for human transplant kidneys should be prohibited, since such markets would lead to some potential vendors being precluded from exercising their autonomy as a result of their ambivalence towards the sale of their organs.

Responses to the first three arguments from irresistible offers

It is clear that the least persuasive of the above three arguments from irresistible offers is the first, for it is highly unlikely that its first premise (that for some potential vendors the price that is offered for their kidneys would be so great as to be literally irresistible) would ever be true. It is implausible to claim that some persons would find the price offered for their kidneys to be so great that they literally could not resist selling their organs against their own wills. Money might sing a siren's song to many, but it is unlikely that this song will have the same effect as that of the original sirens. The second type of argument from irresistible offers can also be readily dismissed. Although the central claim of this argument is likely to be true (that some potential vendors will suffer from compromised autonomy as a result of selling their kidneys through weakness of will) this will be true for markets in *any* good, not just human transplant kidneys. As such, if this argument were to be accepted it would prove too much, for it show that not only should current markets in human transplant kidneys be prohibited, but that current markets in *all* goods should be prohibited. And this is clearly a *reductio ad absurdum* of this anti-market position.

The third argument from irresistible offers, however, cannot be dismissed so quickly. This argument draws on the intuition that were a current market for human transplant kidneys to be introduced, then the range of eligible options that some potential vendors would be faced with would be reduced to one. Since such vendors would no longer be able to exercise their autonomy as fully as they could when they could choose from a range of options, respect for their autonomy requires that one oppose the introduction of such a market. This is certainly plausible. Recognising this, the proponents of current markets in human kidneys have attempted to undermine this anti-market argument by rhetorically asking if one would be more likely to think that such a market was morally acceptable if the appeal of the purchase price for organs was muted. Stephen Wilkinson, for example, notes that the concern for the autonomy of the poor that undergirds this third argument from irresistibility could be met by prohibiting 'the purchase of organs from people below a certain level of wealth', while Gerald Dworkin similarly asks if one would be more inclined to favour allowing organ sales in the lower 40 per cent income bracket were prohibited from participating in the market as organ vendors.[27] Such rhetorical responses to this third anti-market argument are, however, flawed. This anti-market argument is based on the view that the autonomy of the potential vendors is valuable and should be protected. However, the above pro-market responses to this argument do not appeal to intuitions concerning the autonomy of potential vendors, but, instead, to intuitions concerning such vendors' *wellbeing*. These responses rely on those to whom they are addressed believing that all things considered the poor might be *better off* being able to sell their kidneys were they to choose to do so, and for this reason believing that they should be allowed to sell. Such responses thus fail directly to address the *autonomy*-based anti-market argument that they are aimed at. As such, such rhetorical responses to this anti-market argument do not show why this argument is flawed – if, indeed, this argument is flawed at all.

The failure of these rhetorical responses to the third argument from irresistibility to show that it is flawed is unfortunate, for this third anti-market argument *is* flawed. This argument is based on the view that the less eligible options a person has to choose from then the less she will be able to exercise her autonomy (at least once the number of options that she can choose from falls below a certain number).[28] This view of the relationship between a person's ability to exercise her autonomy and her possession of a range of eligible options is plausible owing to the rationalistic overtones of the concept of autonomy, such that a person is exercising his autonomy to a higher degree when he deliberates carefully about the options that are available to him and chooses one on the basis of that deliberation.[29] For persons who accept the rationalistic connotations of the concept of autonomy it makes sense to believe that if a person only has one eligible option available to him (or, more precisely, if he believes that he only has one eligible option available to him) then he will not greatly exercise his autonomy in choosing it. But this view of what is involved in a person exercising his autonomy conflates the *deliberative process* that a person might employ in the exercise of his autonomy with the *exercise* of his autonomy. A person might, for example, simply have no need to deliberate carefully about his potential course of action because the path that he should take is clear to him. This does not imply that such a person acts unthinkingly, as a non-autonomous agent such as a cur dog or a small child might act unthinkingly. Rather, it simply implies that in this instance he found it easy to discover what course of action he should pursue, and so found it easy to exercise his autonomy effectively. Since this is so, the fact that a person finds one particular course of action to be so appealing (such as selling his kidney for a high price) that he does not need to deliberate for long prior to choosing it does not show that he thereby suffers from a compromised ability to exercise his autonomy.[30]

Responses to the fourth argument from irresistible offers

The fourth anti-market argument from irresistibility fares no better than the first three. According to the proponents of this argument persons who are ambivalent about selling their kidneys will thereby be precluded from exercising their autonomy. Since this is so, the proponents of this argument conclude, respect for the autonomy of such potentially conflicted vendors requires that current markets for human kidneys continue to be prohibited. As is stands, however, this argument is incomplete, for its conclusion will only follow if autonomy possesses *intrinsic*, and not merely *instrumental*, value. A person who is ambivalent towards the sale of one of his kidneys is ambivalent because, given his motivational psychology, both the sale of his kidney and its retention are equally attractive to him. Since this is so, when he picks a course of action concerning whether to sell he will not be directed in this picking by his motivational set; his values, desires, pro and con attitudes, and so on. Given that this person is genuinely ambivalent, his motivational set could just equally have justified the *opposite* course of action to that which he picked. Such a person would have no identifiable self to ground his self-direction, and so would be precluded from exercising his autonomy. However, such preclusion would not matter if the ambivalent person's autonomy were only of *instrumental* value to him. If a person's autonomy were only of instrumental value to him it would be of value to him only in those situations when it would matter for him whether it was he, or someone else, who was directing his actions. Were he to be in a situation where it did not matter if it were he or some other directing person his actions, or where it would be preferable from his own point of view for another person to direct his actions, then the exercise of his autonomy would have (respectively) *no* value or *negative* value. And the first of these two situations is that which the genuinely ambivalent potential kidney vendor finds himself in. Given that such a potential vendor has no reason to choose to sell his kidney over choosing to retain it, it does not matter from the point of view of one concerned solely with securing the best outcome for him (i.e. from the point of view of one who is concerned solely with the instrumental value of his autonomy) whether it is the potential vendor, or someone else, who makes the decision as to whether or not he should sell. The instrumental value of the autonomy of such a potential vendor is thus zero. If autonomy were of only instrumental value, then, it would not matter from the point of view of one who valued autonomy highly that the ambivalent potential kidney seller would be precluded from exercising his autonomy when faced with the choice of selling or retaining one of his kidneys. Given the motivational set and epistemic limitations of such a potential vendor his autonomy was of no instrumental value to him in this situation anyway.

If one believes that autonomy is only of instrumental value, then, this fourth type of argument from irresistibility has no force. Thus, as I noted above, for this argument to have force it must rest on the view that autonomy is *intrinsically* valuable. And the more intrinsically valuable one believes autonomy to be the stronger will be this fourth anti-market argument from irresistibility, for the more intrinsically valuable autonomy is the more reason we have to prohibit that which adversely affects it. To complete this fourth argument from irresistibility, then, one must show that autonomy (as it is understood in the context of this discussion) is intrinsically valuable. Moreover, for this fourth argument from irresistibility to be a strong argument against allowing markets for human kidneys one must show that autonomy is of high intrinsic value. Conversely, to rebut this anti-market argument one must demonstrate that autonomy is *not* intrinsically valuable in its own right. Alternatively, if one believed that autonomy *is* intrinsically valuable, one could attempt to weaken this fourth anti-market argument from irresistibility by demonstrating that the intrinsic value of autonomy is not high.

Although a discussion of the way in which autonomy should be valued is beyond the scope of this chapter there are two brief arguments that can be used to support the above two claims concerning the value of autonomy that must be demonstrated to rebut the above argument from irresistibility. Consider here the argument that Dworkin offers against the view that more choice is better than less:

> Suppose someone ranks three goods A, B, and C in that order. Then, making certain plausible assumptions about the infinite divisibility of utility, there will be A, B, and C such that the person prefers a choice between B and C to receiving A. This will occur when the utility of having a choice between B and C plus the utility of B is greater than the utility of A. This seems to me irrational. Leaving aside some special feature about this particular choice, for example, that somebody promised me $1,000 if I made the choice between B and C, why should I prefer to receive my second-ranked alternative to my first?[31]

Putting Dworkin's argument in terms of the value of exercising one's autonomy rather than in terms of the value of making a choice, the value to rational persons of being able to exercise their autonomy by choosing between B and C would be less than the value to them of the difference in value, D, that they believe holds between A and B. However, that the value of D would be held by rational persons to be greater than the value of the exercise of their autonomy in choosing between B and C does not show that the exercise of their autonomy is not intrinsically valuable. Instead, this just shows that the intrinsic value ascribed to the exercise of their autonomy is less than the value ascribed to D. Yet since Dworkin's argument, as stated, does not rest on any minimum value being ascribed to D this reformulation of Dworkin's argument shows that the intrinsic value of autonomy, if any, is likely to be low, for it is likely that rational people would choose A even when the value of D is low. Since this is so, this reformulation of Dworkin's argument supports the second of the above claims that must be demonstrated to rebut the fourth anti-market argument from irresistibility. If persons' exercise of their autonomy is only of low intrinsic value (as it seems to be), then even if markets in human transplant kidneys would preclude persons from exercising it this would provide only a weak reason for their prohibition.[32]

Shifting the focus from autonomy to wellbeing

Yet although Dworkin does not believe that his argument above depends on any minimum value being ascribed to D one could still object to it on the grounds that for some range of values ascribed to D it would be rational for some persons to prefer to choose between B and C rather than just to be provided with A. Such persons would be those who prefer to make their own choices and direct their own lives, even when they know that they would be more likely to achieve their goals (other than the goal of exercising their autonomy) were another paternalistically to make their choices for them. For such persons, then, until the value of D exceeded the value that they placed on the exercise of their autonomy in making their own choices it would be rational for them to choose between B and C even though they realise that this would preclude them from securing A. Moreover, this preference for the exercise of autonomy seems to be widely held. Persons frequently resent others who paternalistically interfere with their lives, even if they recognise that such interference might be of benefit to them. At first sight, this seems to support the fourth anti-market argument from irresistibility. If the exercise of autonomy is widely held to be intrinsically valuable to some degree, and if allowing markets in human transplant organs would preclude a subgroup of potential vendors from exercising their autonomy, then it seems that out of respect for their autonomy one should support the continued prohibition of such markets.

Yet this anti-market conclusion comes too quickly. Persons who prefer to exercise their autonomy in making their own choices often prefer to do so because they believe that, given their preferences, they would be better off doing so rather than being subject to the paternalistic direction of another, even if they recognise that such paternalistic direction might be more likely to secure for them their non-autonomy-based goals. For such persons the exercise of their autonomy is not merely a *means to* their wellbeing, but *part of* it. This is important for two reasons. First, if the value of autonomy is derivative from that of wellbeing, then this supports the first of the claims outlined above that must be demonstrated to rebut the fourth anti-market argument from irresistibility (i.e. that autonomy is not intrinsically valuable in its own right). Second, and in a related vein, once it is that the value of autonomy is derivative in this way the focus of this fourth ambivalence-based objection to current markets in human transplant kidneys shifts from the question of whether such markets would preclude some vendors from exercising their autonomy to the question of whether such markets would adversely affect the wellbeing of potential vendors. Moreover, since those potential vendors who would be ambivalent about the sale of their organs are not morally privileged in any way, the degree to which their wellbeing would be affected by allowing current markets in human organs must be weighed against the wellbeing of other groups of persons who would also be affected by removing the prohibitions on such markets. Absent any definitive argument for the intrinsic value of autonomy that is not derivative from the wellbeing of its possessors, then, this fourth type of anti-market argument from irresistibility leads not to a condemnation of current markets for human transplant kidneys, but, instead, to a need to assess such markets in the light of the effects that they would be likely to have on human wellbeing.

Wellbeing and kidney sales

Despite the claims of their proponents, the autonomy-based anti-market arguments discussed above do not support the conclusion that respect for autonomy supports the continued prohibition of current markets in human transplant organs. Instead, they are either fatally flawed or lead to the conclusion that the moral permissibility (or otherwise) of such markets rests on the effects that such markets would have on human wellbeing. This change in focus, however, might not be an unwelcome one for those who hold that current markets in human transplant organs are morally impermissible. The most extensive empirical study of the effects that a current market in human kidneys has on the wellbeing of those who participated in it as vendors has show that the wellbeing of such persons is severely compromised as a result of their selling their kidneys. This study was conducted by Madhav Goyal et al. in February 2001, and focused on the effects that the sale of a kidney had on 305 persons who had sold their kidneys in Chennai, a city in the state of Tamil Nadu in southern India.[33] Many of the kidney vendors who were interviewed by Goyal et al. reported a worsening of their economic status following the sale of their kidneys, with the 'average family income ... [declining] ... from $660 at the time of nephrectomy to $420 at the time of the survey, a decrease of one third.' (In Tamil Nadu the poverty line for an average-sized family is $538.) This decline in income was not, however, offset by the vendors having been able to accumulate capital through the sale of their kidney. Indeed, only 11 per cent of the kidney vendors that Goyal et al. surveyed were able to retain any of the money that they received from the sale of their kidneys as either cash, or the equivalent of cash (such as jewellery or investments). The majority of the vendors (60 per cent) had to disburse the money that they received to pay off debts, or to buy food (22 per cent). Worse yet, the decrease in the family income that these vendors reported occurred at a time when the economic situation of the poor of Tamil Nadu was improving. Over the last five years per capita income in Tamil Nadu has increased by 10 per cent, and over the last ten years by 37 per cent (both figures are after adjusting for inflation), and 'the proportion of people living below the poverty line has declined by more than 50 per cent since 1988'. Moreover, in addition to their economic decline, the kidney vendors of Chennai suffered from a decline in their health as a consequence of the sale of their kidneys. When they were asked to rank their health status before and after the sale of their kidney using a five-point Likert scale (a standard scale used to measure persons' attitudes) which ranged from 'excellent' to 'poor', 38 per cent of the vendors reported a one- to two-point decline, and 48 per cent reported a three- to four-point decline. A third of the kidney vendors surveyed also reported long-term back pain as a result of their nephrectomy, and half reported that they suffered from persistent pain at the site of the nephrectomy.[34]

From these figures it might seem that the sale of a kidney in a current market for such would lead to a reduction in both the economic status and the healthy status of the kidney vendors. Thus, if one were concerned about the effects on wellbeing that such a market would have on those who would participate in it as vendors (as the fourth autonomy-based argument, above, leads one to be), then one should support its continued prohibition. Yet to draw this conclusion would be too hasty. Kidney sales are banned in India under the Transplantation of Human Organs Act of 1994, which prohibits any form of trafficking in human organs. The kidney vendors that Goyal et al. surveyed in Chennai thus sold their kidneys illegally, on the black market. As such, they had no legal recourse against the immediate purchasers of their kidneys if such purchasers defrauded them by failing to compensate them at the agreed level, or if such purchasers failed to provide the post-operative care that they promised. And the survey conducted by Goyal et al. showed that fraud was endemic in the black market, with the typical vendor receiving on average payment that was one third less than promised.[35] Moreover, as well as being subject to fraud, the Chennai vendors failed to receive adequate medical care. Indeed, reports are legion of kidney vendors in Chennai simply being dumped back on the streets after the removal of their kidneys.[36]

The adverse economic and health consequences of selling a kidney in India can thus be traced to the vendors in this illegal market being defrauded, and to the inadequate medical care that they received after selling their kidneys. (It is also likely that the vendors are in poor physical condition at the time of the sale, and this exacerbates their post-operative poor health.) But fraud and poor medical care are not necessary features of a current market for human transplant kidneys. They are merely the likely results of the illegality of such a market. Were a current market in human transplant kidneys to be legalised and regulated persons who would participate in it as vendors would not be as likely to be defrauded, for they would have legal recourse against those to whom they sold their organs. Moreover, were such a market to be regulated, the direct purchasers of the kidneys sold within it could be required only to purchase organs from persons who met certain minimal health conditions. They could also be required to provide them with adequate post-operative care, such as a paid 4-12-week recuperation period, extended follow-up examinations (and any treatment judged necessary as a result) and adequate analgesia.[37] The legalisation and regulation of a market for human transplant kidneys would thus help to eliminate the adverse economic and health consequences that currently plague the kidney vendors in Chennai's illegal kidney market. Of course, that such adverse consequences of the sale of a kidney would largely disappear in a legal and regulated market for human transplant kidneys does not show that allowing persons to sell their kidneys in such a market would necessarily enhance their wellbeing. The vendors might suffer from unexpected complications as a result of surgery or from 'seller's remorse' once the transaction had taken place. Such possibilities reinforce the need for strict oversight of any current market in which persons sell their kidneys, both to minimise the possibility of medical complications and also to ensure that the vendors give their informed consent to the sale of their kidneys. With such safeguards in place those who still wished to sell their kidneys would desire this as they believed that, given their personal values and desires, such sales would most likely enhance their wellbeing as they understand it. Since this is so, then, *ceteris paribus*, the presumption should be that such potential vendors are the best judges of what would serve their own interests. If one is concerned with human wellbeing, then one should support, rather than oppose, the view that a legalised, regulated current market for human kidneys is morally permissible.

Autonomy, poverty and welfare

It seems, then, that concern for human wellbeing should lead one to hold that a legal, regulated market in human transplant kidneys is morally permissible. Moreover, given the failure of both the anti-market argument from economic coercion and the four anti-market arguments from irresistibility to demonstrate that the autonomy-based pro-market argument outlined above is mistaken, it appears that respect for the autonomy of those who would participate in such a market as vendors also supports the view that such a market is morally permissible. Yet even with these arguments in hand one might still suspect that, from the point of view of one who values autonomy, there is something illicit about a current market for human organs. One might, for example, still believe that the poor are more subject to being forced to perform certain actions (such as selling their kidneys) owing to their poverty in a way that their wealthier brethren are immune from. If so, then *despite* the arguments above one might still believe that the poor would somehow be forced into selling their kidneys were a current market for them be legalised, and so would suffer from a diminution in their autonomy with respect to their vending actions. Alternatively, one might think that – again, despite the arguments above – concern for the wellbeing of the poor should not lead one to advocate allowing them to sell their kidneys if they so wished, but, instead, to advocate some other means of lifting them out of poverty and improving their lot.

Both these intuitions are widely held. Moreover, both seem to support the view that something has gone wrong in the above pro-market arguments, and that, once their flaws have been detected, it will be a simple matter to show why they should be rejected. Instead of trying to defend the above pro-market arguments by arguing that these common intuitions are not well founded I will defend them by arguing that although these intuitions *are* well founded, the conclusions that persons often draw from them are misguided.

I will begin by addressing the intuition that, as a result of their poverty, the poor are somehow forced into performing actions that they would not otherwise wish to perform, and that this renders them less autonomous with respect to such actions. This intuition is a plausible one. After all, it seems wrong to claim that an impoverished person in the slums of Chennai is just able to direct his own life in accord with his desires and values, is just as autonomous as (for example) an affluent Englishman living in a leafy suburb of London. But care must be taken with the claim that the Chennai slum-dweller is less autonomous that the affluent inhabitant of a London suburb, for this claim is ambiguous. On the one hand, this might be understood as the claim that the Chennai slum-dweller has a lower *capacity* for autonomy than the affluent Englishman. On the other hand, this might be understood as the claim that the Chennai slum-dweller is less able to *exercise* his autonomy so as to act in accord with his own desires and values. The first way of understanding this claim is simply false, for there is no necessary correlation between a person's wealth and his capacity to make decisions on the basis of his desires and values. Yet the second way of understanding this claim is true, for owing to his greater wealth it is more likely that the Englishmen would be able to exercise his autonomy to realise his goals. However, this does not mean that the affluent Englishman is *more autonomous* that the impoverished Indian. Rather, all this means is that the Englishman's autonomy is more *instrumentally valuable* to him that the Indian's autonomy is to him. Accordingly, to claim that a person is forced into performing a certain action by his poverty is not to claim that he suffers from compromised autonomy with respect to that action, but instead to claim that his autonomy is not as instrumentally valuable for him as it might have been. But once this is recognised it is clear (from the argument concerning wellbeing, above) that the instrumental value of the poor potential organ vendors would be enhanced, not diminished, were they allowed to sell their kidneys in a regulated current market for them, for they would then be able to use their autonomy to trade their kidneys for something that they valued more. Thus, the intuition that the poor are somehow forced by their poverty into performing actions that they would otherwise not wish to perform supports, rather than undercuts, the view that current markets in human kidneys should be allowed.

This analysis of the common intuition that poverty compromises autonomy leads directly to the intuition that the correct moral response to poverty is not to allow persons to sell their kidneys in an attempt to improve their lot, but instead to provide them with aid so that they no longer have to pursue such a desperate remedy. Unlike the above intuition concerning the autonomy-undermining effects of poverty, this intuition does not need to be unpacked. However, one can accept this intuition (that is, one can accept that persons have a moral duty to aid the poor) and yet *still* hold current markets in human transplant kidneys to be morally permissible. It is not true (as many critics of the market appear to believe) that if one holds that a current market in human transplant kidneys is morally permissible, one must also hold that allowing such a market will suffice to discharge one's duty of aid to the poor. Indeed, the only case in which one could not accept the intuition that one has a duty to aid the poor while also holding that current markets in human kidneys are morally permissible would be when one held that such markets are morally permissible *only if* they secure as many transplant kidneys as possible. Such an extreme pro-procurement position is inconsistent with the intuition that one has a moral duty to aid the poor as if one acts on this duty and improves the economic lot of the poor one is likely to reduce the number of persons who would wish to sell their kidneys. But this extreme pro-procurement position is not that which I defend here. Instead, I am merely defending the claim that persons should, if they so wish, be allowed to sell their kidneys in a legal, regulated, current market for them. And this moderate pro-market position is fully consistent with accepting the view that there is a moral duty to lift the poor out of their poverty. Indeed, to the extent that one holds that such a market is morally required because one believes that it will help to improve the lot of the poor one is more likely to accept that one has a moral duty to aid them than not.

Conclusion

It is now time to take stock. I have argued above that rather than militating against allowing persons to sell their kidneys in a current market, respect for the autonomy and concern for the wellbeing of the potential vendors should move one to support such a market. This is not, however, a definitive defence of such a market. Indeed, it is not even a definitive defence of such a market in the eyes of one who believe that autonomy should be the primary value of medical ethics. First, I have focused here only on the effects that such a market would have on the autonomy of the potential vendors, and have not addressed the issue of the effects that it would have on the autonomy of the potential recipients of the kidneys that they would sell. And if, as many critics of such a market allege, the organs procured in a current market would be of poor quality and so would jeopardise the lives of those who received them, it might be the case that allowing such a market would compromise the autonomy of the recipients, even if it would not compromise the autonomy of the vendors. If so, then from the point of view of one who holds respect for autonomy to be an important moral value this would militate against the view that such a market is morally permissible. Second, it is possible that irrespective of the effects that a market in human kidneys would have on the autonomy of those who would participate in it, it is morally impermissible to commodify human body parts in the way that such a market would require. Third, it might be that the sale of a kidney is so dangerous that the paternalistic prohibition of such sales is warranted. Finally, it might be that even if it can be shown that a current market for human kidneys is morally permissible, that would be good public policy reasons not to legalise it. It might, for example, be likely that such a market would be so prone to abuse that even though it is not morally impermissible *per se* to buy and sell human kidneys, the disadvantages of allowing such a market in practice would greatly outweigh its advantages.[38] For the purposes of this chapter, then, it is still an open question as to whether a current market for human kidneys is morally permissible. However, the widespread objection that it would serve to compromise the autonomy of those who would participate in it as vendors through subjecting them to coercion or irresistible inducements is mistaken.[39]

Notes

1 Janet Radcliffe Richards (1996) 'Nepharious Goings On: Kidney Sales and Moral Arguments', *The Journal of Medicine and Philosophy*, 21, p. 375.
2 Leon Kass (2002) *Life, Liberty and the Defense of Dignity* (San Francisco: Encounter Books), pp. 177–98; Nancy Scheper-Hughes (2003) 'Keeping an Eye on the Global Traffic in Human Organs', *The Lancet*, 361 (10 May), pp. 1645–8; Lawrence Cohen (1999) 'Where it Hurts: Indian Material for an Ethics of Organ Transplantation', *Daedalus*, 128(4), pp. 135–65.

3 See, for example, Mario Morelli (1999) 'Commerce in Organs: A Kantian Critique', *Journal of Social Philosophy*, 30(2), pp. 315–24; and T. L. Zutlevics (2001) 'Markets and the Needy: Organ Sales or Aid?, *Journal of Applied Philosophy*, 18(3), pp. 297–302.

4 See Janet Smith (1997) 'The Pre-Eminence of Autonomy in Bioethics', in David S. Oderberg and Jacqueline A. Laing (eds.) *Human Lives: Critical Essays on Consequentialist Bioethics* (New York: St. Martin's Press), pp. 182–95.

5 This autonomy-based pro-market argument is offered by Gerald Dworkin (1994) 'Markets and Morals: The Case for Organ Sales', in Gerald Dworkin (ed.), *Morality, Harm, and the Law* (Boulder, CO: Westview Press), pp. 155–61.

6 See James Stacey Taylor (2002) 'Autonomy, Constraining Options, and Organ Sales', *Journal of Applied Philosophy*, 19(3), pp. 273–85; and (2004) *Stakes and Kidneys: Why Markets in Human Body Parts are Morally Imperative* (Aldershot: Ashgate Press), chapter 4. The two most prominent proponents of the view that the sale of an organ is an autonomy-compromising constraining option are Paul Hughes (1998) 'Exploitation, Autonomy, and the Case for Organ Sales', *International Journal of Applied Philosophy*, 12.1, pp. 89–95, and Zutlevics, 'Markets and the Needy'.

7 I have previously defended the first of these premises in *Stakes and Kidneys*, chapters 2 and 3. However, my focus in that volume was different from my focus here, for there I concentrated on determining whether Dworkin's views on coercion as expressed in Gerald Dworkin (1970) 'Acting Freely', *Nous*, 4(4), pp. 367–85, committed him to holding that the reluctant kidney vendor was coerced into selling.

8 Noted by Radcliffe Richards, 'Nepharious Goings On', p. 376.

9 Erin and Harris argue that rather than using a market to distribute the organs thus procured, a single buyer such as the NHS should purchase them and distribute them according to non-market criteria. See Charles A. Erin and John Harris (2003) 'An Ethical Market in Human Organs', *Journal of Medical Ethics*, 29(3), pp. 137–8. I argue against this proposal and in favour of a regulated market for the distribution of in human organs in *Stakes and Kidneys*, chapters 5 and 9.

10 Harry G. Frankfurt (1988) 'Freedom of the Will and the Concept of a Person', in Harry G. Frankfurt (ed.), *The Importance of What We Care About* (Cambridge: Cambridge University Press), p. 11.

11 Frankfurt, 'Freedom of the Will', p. 12.

12 Frankfurt, 'Freedom of the Will', p. 19.

13 Frankfurt, 'Freedom of the Will', p. 20.

14 For Hobbes, 'the last appetite, or aversion, immediately adhering to the action, or to the omission thereof, is that we call the will'. Thomas Hobbes (1994) *Leviathan*, ed. Edwin Curley (Indianapolis, IN: Hackett Publishing), Part 1, chapter 6, para. 21. See Frankfurt, 'Freedom of the Will', p. 14.

15 Frankfurt, 'Freedom of the Will', p. 20.

16 For an outline of such criticisms see my Introduction to James Stacey Taylor (ed.) (2005) *Personal Autonomy: New Essays in Personal Autonomy and its Role in Contemporary Moral Philosophy* (Cambridge: Cambridge University Press), pp. 4–10.

17 The most recent of these modifications and clarifications can be found in Frankfurt's replies to his critics, in Sarah Buss and Lee Overton (eds.) (2002)

The Contours of Agency: Essays on Themes from Harry Frankfurt (Cambridge, MA: MIT Press).

18 See, for example, John Christman, Introduction to John Christman (ed.) (1988) *The Inner Citadel: Essays on Individual Autonomy* (Oxford: Oxford University Press), pp. 8–9. See also James Stacey Taylor, Introduction," in Taylor (ed.), *Personal Autonomy*, p. 4.

19 That is, Frankfurt makes no mention of autonomy when he discusses what it is for a person to identify with his effective first-order desires. He does, however, write of what it is for a person to be autonomous with respect to his actions. See his 'Coercion and Moral Responsibility', in Frankfurt (ed.), *The Importance of What We Care About*, p. 43. This, however, is compatible with the claim I make above, since for Frankfurt that a person identifies with his effective first-order desire is necessary but not sufficient for him to be autonomous with respect to the action that it moves him to perform. I argue for this construal of Frankfurt's view in James Stacey Taylor (2003) 'Autonomy, Duress, and Coercion', *Social Philosophy & Policy*, 20(2), pp. 127–55 – although there I mistakenly hold that a person's identifying with his effective first-order desire and his being autonomous with respect to it are synonyms.

20 The first of these mistakes can be found in Ann Cunningham (2003) 'Autonomous Consumption: Buying into the Ideology of Capitalism', *Journal of Business Ethics*, 48(3), pp. 229–36. Here, Cunningham draws on Noggle's analysis of what it is for a person to identify with a desire (i.e. of what it is for a person to fail to be alienated from a desire) to defend advertising from the change that it undermines personal autonomy. In fairness to Cunningham, however, Noggle also believes that autonomy and identification are coextensive. See Robert Noggle (1995) 'Autonomy, Value and Conditioned Desire', *American Philosophical Quarterly*, 32(1), pp. 57–69. The second of these mistakes is made by Nomy Arpaly, 'Responsibility, Applied Ethics, and Complex Autonomy Theories'," in Taylor (ed.), *Personal Autonomy*, pp. 173–5.

21 Versions of this argument have been offered by Pranlal Manga (1987) 'A Commercial Market for Organs? Why Not', *Bioethics*, 1(4), p. 327; John B. Dossetor and V. Manickavel (1992) 'Commercialization: The Buying and Selling of Kidneys', in C. M. Kjellstrand and J. B. Dossetor (eds.), *Ethical Problems in Dialysis and Transplantation* (Dordrecht: Kluwer Academic Publishers), p. 63; and Patricia A. Marshall, David C. Thomasma and A. S. Daar (1996) 'Marketing Human Organs: The Autonomy Paradox', *Theoretical Medicine*, 17, p. 13. I have addressed this argument at length in *Stakes and Kidneys*, chapter 3.

22 All parties to this debate accept that the typical vendor will be poor.

23 One might respond to this objection to the anti-market argument from economic coercion by arguing that if persons intentionally act so as to keep the poor in such poverty that the sale of their organs becomes (or remains) their best option, then the persons who propagate this state of affairs would be manipulating the poor in such a way as to compromise their autonomy with respect to their vending actions. A version of this argument has been offered by Zutlevics, 'Markets and the Needy', pp. 297–302. For a response to it see my *Stakes and Kidneys*, chapter 4. I thank Nafsika Athanassoulis for pressing me on this point.

24 Sells writes that since 'the financial benefits [of selling a kidney] [would] have such an impact on the life of the donor ... as to be irresistible: the element of voluntariness ... must be ... in extreme cases, abolished'. R. A. Sells (1991) 'Voluntarism of Consent', in W. Land and J. Dossestor (eds.), *Organ Replacement Therapy: Ethics, Justice, Commerce* (New York: Springer-Verlag), p. 20. I address Sells' argument more fully and directly in *Stakes and Kidneys*, 67–9.

25 I thank Thomas Magnell for bringing this objection to my attention.

26 See Paul Hughes, 'Autonomy, Ambivalence, and Organ Sales', unpublished MS. A version of this argument is also developed in Ruth Grant and Jeremy Sugarman (2004) 'Ethics in Human Subjects Research: Do Incentives Matter?', *Journal of Medicine and Philosophy*, 29, pp. 717–38.

27 Stephen Wilkinson (2003), *Bodies for Sale: Ethics and Exploitation in the Human Body Trade* (London: Routledge), p. 131. In fairness to Wilkinson it should be noted that he is not here directly addressing the third anti-market argument from irresistibility, but instead the related charge that markets for human organs are exploitative. Dworkin, 'Markets and Morals', p. 157.

28 The adherents of this view are not committed to the view that the more options a person has the more able he is to exercise his autonomy (although some of them might endorse this). Nor are they committed to the view that a person must have a range of options available to her (although, again, some of them might endorse this). Instead, they are only committed to the weaker view that she must believe that she has a range of options open to her.

29 This view is also plausible as it is plausible to believe that a person is less autonomous, less able to direct his own life in accordance with his desires and his values, if he has a low number of eligible options. I address this below, in the section entitled 'Autonomy, Poverty, and Welfare'.

30 Indeed, if the course of action that a person should take is clear to him, owing either to the unthinkability of the alternatives, or to this course of action being required by his volitional nature, then it is plausible to claim that the person concerned will exercise his autonomy *most* fully when he pursues it. See Harry G. Frankfurt (1999) 'Autonomy, Necessity, and Love', in Harry G. Frankfurt, *Necessity, Volition, and Love* (Cambridge: Cambridge University Press), pp. 129–41.

31 Gerald Dworkin (1988) *The Theory and Practice of Autonomy* (Cambridge: Cambridge University Press), p. 80.

32 One might object that this argument only shows that the *exercise* of autonomy is of low intrinsic value (if it is of intrinsic value at all), and not that autonomy *per se* is of low intrinsic value. This is certainly correct. However, since it is not clear how autonomy could possess value independent of its exercise (or its potential for exercise) showing that the exercise of autonomy is of low intrinsic value (if it is of any intrinsic value at all) *de facto* demonstrates the low (or absent) intrinsic value of autonomy *per se*.

33 Madhav Goyal et al. (2002) 'Economic and Health Consequences of Selling a Kidney in India', *Journal of the American Medical Association*, 288(13), pp. 1589–93. The findings of Goyal et al. are also outlined in Taylor, *Stakes and Kidneys*, pp. 77–8, 84–5.

34 Goyal et al., 'Economic and Health Consequences of Selling a Kidney in India', pp. 1590–2.

35 Goyal et al., 'Economic and Health Consequences of Selling a Kidney in India', p. 1591.

36 Praveen Swami (2003) 'Punjab's Kidney Industry', *Frontline*, 20(3), 1–4 February.

37 Such medical care for persons who undergo nephrectomies is suggested by Working Party of the British Transplantation Society and the Renal Association, *United Kingdom Guidelines for Living Kidney Donor Transplantation* (2000). See also Taylor, *Stakes and Kidneys*, pp. 87–8, where I develop this pro-market argument further.

38 I argue against all of these further objections to markets in human kidneys in *Stakes and Kidneys*, chapters 6, 7 and 8.

39 An earlier version of this chapter was presented at the Murphy Institute, Tulane University, in January 2005. I thank my audience on that occasion (especially Eric Mack and Bill Glod) for their helpful comments. I also thank Nafsika Athanassoulis for her useful comments on an earlier version.

7
The Role of Conscience in Medical Ethics

Piers Benn

There are a number of areas of applied philosophy where the idea of conscience enjoys pivotal importance. In political philosophy, one may debate whether elected representatives, say in Parliament, have a duty first and foremost to vote in accordance with their individual conscience, or the manifesto they were elected on, should there be a conflict; in warfare it is debated whether soldiers should follow their conscience in disobeying orders they consider to be immoral; in business it is debated whether employees' primary loyalty should be to their employers or to their conscience, in cases where employees suspect abuses are taking place. But, at least in the public mind, it is probably in health care that issues of conscience are most clearly apparent. Examples of doctors or nurses who refuse to assist in abortion, because it goes against their conscience, or of health care professionals who break the law on euthanasia or assisted suicide because they feel prompted to do so by their conscience, are all too easy to find. But what exactly is the appeal to conscience, and what authority should it have? When, if ever, should provisions for conscientious objection be made? In this chapter I shall explore the logical structure of appeals to conscience, and then ask how appeals to conscience are best analysed in medicine and nursing.

Conscience and moral values

It is often thought that following one's conscience is noble and upright, particularly when there is a serious personal cost in doing so. Many people who are not themselves pacifists look with admiration on conscientious objectors during the First and even the Second World War, especially in view of the harsh treatment they often received –

the accusations of cowardice, of sympathy with the enemy, of shirking their duty to their country. When ministers resign their office because they cannot agree with central policies of the government they are meant to be serving, their esteem in the eyes of the public often rises. All of this leads to the question of whether there really is anything admirable about following one's conscience, over and above simply doing the right thing. As we shall see, questions arise about how we should morally assess people who, in good conscience, do wrong or even terrible things.

A number of claims have been made about conscience. First, it has been held at least since St Thomas Aquinas that acting according to one's conscience is not enough to exonerate one from blame, if one does what is objectively wrong. People can be prompted by their conscience to do terrible things, and it is plausible to think that their conviction of their own righteousness makes them worse rather than better. Some such people are called fanatics, and this is no compliment. To spend years planning and then carrying out a major terrorist atrocity against innocent civilians in the unwavering conviction that one is doing something beautiful and good, and even divinely ordained, is the mark of fanatical evildoer. At the same time, Aquinas and many subsequent thinkers hold that it is still necessary to act in accordance with one's conscience if one is to escape moral blame. To do something that one thinks is wrong is itself wrong, even if the act is actually permissible.

According to the Thomist account, then, permissible action requires two conditions: the agent must act in a way that is objectively permissible, and must also believe that the action is permissible. If either condition is lacking, the agent acts wrongly. Acting according to one's conscience is necessary but not sufficient for acting in a morally permissible way.[1]

Second, the appeal to conscience seems to have certain features that set it apart from simple appeal to moral values. For example, although other people can act in ways I morally disapprove of, they cannot strictly speaking act against my conscience. Of course, they can try to persuade me to act in a way I think wrong, or they can try to make me complicit in some wrong act. But if I succumb, then it is I, and not they, who properly speaking act against my conscience.

This feature of conscience can give a misleading impression that there is a code of ethics that applies only to me – a sort of moral solipsism. Appeals to personal conscience are indeed self-addressed – they concern my conduct, not the conduct of others. But this does not mean that moral requirements that apply to me do not also apply to others. We should not be misled by such commonplace exhortations as 'do as your conscience dictates, but allow me to do as mine dictates', or 'my conscience tells me to do X, but I don't judge anyone else – this is my personal conviction'. These familiar utterances lend no real support to the idea that different things are right for different people in otherwise similar circumstances. Nor do they lend support to any form of moral relativism. Properly understood, such utterances reflect the idea that it is wrong for anyone to do what he thinks is wrong. But at the same time, one may have false views of what is right and wrong. To the requirement that each should follow his or her conscience should be added that each should acquire a conscience that is as morally sound as possible.

A third feature of conscience is its alleged role as a source of moral knowledge or wisdom. Historically, the rise of Protestantism is largely responsible for this, with its implicit individualism and opposition to priestly authority and certain kinds of church hierarchies. There is a strong strand of opposition to following authority (or at least, following it blindly – note the subtle but crucial difference) and a corresponding conviction that the individual conscience is the proper source of moral knowledge. It is the remnants of this conviction that explains the admiration many people feel for those who disobey orders they consider to be wrong, or speak out against ideas to which their official role – as a Cabinet minister, say – is supposed to commit them. Not only is it thought admirable in itself to follow one's conscience when it conflicts with official authorities, but individual conscience is in itself a more reliable moral guide than authority, broadly understood. This epistemological claim is worthy of some attention.

One approach may be called the 'oracular' conception of conscience. One sometimes hears about people who went away and 'consulted their conscience' to solve some difficult moral conundrum. The phrase suggests that consulting one's conscience is rather like consulting the Delphic Oracle, returning afterwards with vouchsafed knowledge. Yet there is less to this than meets the eye. What, one might ask, is the difference between consulting one's conscience and consulting oneself? And what can consulting oneself really be, if we are to avoid the air of circularity it suggests? Surely it is a matter of thinking hard about the issue, clarifying in one's own mind what one's basic values are, whether those values are right and how they apply to the matter in question. In other words, it is critical and open-minded reflection, with a view to forming a reasonably reliable, though obviously not infallible judgement.

Most of us do this, and if we wish to call it 'consulting our conscience', then so be it. But we should be wary of the danger of underestimating the fallibility of our moral reflections, and also the subtler danger of a certain individualism, which inflates the reliability of one's moral judgement *because* it is one's own. It is a tautology that any moral judgement I make is *my* judgement, and that I cannot disagree with my own judgements as I can disagree with those of others (though obviously I can change my mind). But this clearly does not mean that my capacity for moral reflection is intrinsically better than anyone else's. Indeed, in both morality and elsewhere, the realisation that many people of intelligence and goodwill disagree with us, and that there are many non-rational influences on our most cherished convictions, ought rationally to dislodge at least some of our certainties. We must be prepared to revise and educate the deliverances of conscience.

Much of this is familiar. What of its application to practical ethical issues, for example in medicine? It is helpful first to analyse the claim that it is morally wrong to do something that one considers to be morally wrong. If we can make sense of this, we can explore the interesting suggestion that a person may be wronged or harmed if forced or put under pressure to do something he thinks is wrong. If this holds water, we have a case for respecting the conscience of others and allowing health care professionals, in certain circumstances, the right of conscientious refusal to participate in certain medical procedures. That there should be such a right is widely agreed, but the exact grounds for this are often less clear.

So first, is it true that it is morally wrong to do something that one holds to be morally wrong? We are clearly not talking about merely acting in a way that one knows is forbidden by some authority, or some legal or professional code. In those cases, one might just have a moral disagreement with the authority in question. One might in a loose way admit that what one is doing is wrong, whilst really employing what Hare called the 'inverted commas' sense of 'wrong', exemplified by such thoughts as 'Doctor-assisted suicide is generally considered wrong, but I don't agree, so I'm going to ignore it'. But the cases that concern us here are actions that go against the agent's *own* ethical standards. If such actions are indeed wrong, then doing them with a clear conscience does not exonerate one from blame.

But if one does something that is not actually wrong, but in the belief that it is, what is there to object to? One may respond by saying that such a person shows a *willingness* to do wrong. On this account, such a person shows insufficient concern for morality, and hence may not be trusted to refrain from actions that really are wrong. 'This may be immoral, but so what?' suggests that moral considerations, whether correctly held or not, do not have sufficient weight for the agent. Perhaps the point has more resonance in the traditional theological context in which it was originally made: as Anthony Kenny explains, 'For Aquinas, unlike Kant, the human conscience was not a law-giver. Rather, a man's conscience was his opinion, true or false, about the law made by God. To act against one's conscience was always wrong, because it involved acting against what one believed to be the law of God'.[2]

However, and especially in a secular context, there is an important ambiguity in the phrase 'shows a willingness to do wrong'. Understood one way, it means 'willing to do what is, in fact, wrong', whereas on another interpretation, it means 'willing to do what one believes to be wrong'. As to the first interpretation, it is clearly false that one who does what he *thinks* is wrong shows that he is prepared to do what is *really* wrong. It is logically conceivable that someone who acts against a badly educated conscience rarely does what is really wrong, but instead leads a morally unblemished life while constantly troubled by a guilty conscience. But on the second interpretation we only have a re-statement of the situation we are enquiring about. We still need to know *why* it should be wrong to do something one considers to be wrong.

Knowing one's own moral views

The matter is further complicated by certain speculations concerning the psychology of moral conviction. What, psychologically, is to count as holding a moral conviction? Those who maintain an internalist account of the relationship between moral commitment and inten-tion-formation say that if one holds a moral conviction on a matter relevant to one's own behaviour, then it is a conceptual truth that one will tend to act in accordance with that conviction. For example, R. M. Hare thought this was implied by his prescriptivism. If moral language is primarily prescriptive, and descriptive only in a secondary sense, then assent to moral judgements is not primarily a matter of forming beliefs, but of being disposed to act in accordance with those judge-ments when the occasion arises. Hare inferred from this that the phe-nomenon often called *akrasia* or weakness of will was mis-described – *akrasia* did not really exist, and the contrary appearance was explained either in terms of lack of moral sincerity, or psychological inability to act as one thinks one ought.[3]

However, Hare's account of the matter is overstretched; there are no obvious conceptual obstacles to genuinely *akratic* behaviour. We do not need to deny that there is some important connection between moral judgement and behavioural motivation; all we need question is whether the moral conviction must always be motivationally overriding, when it competes with a non-moral practical consideration. Hare later made it one of the defining features of a moral conviction that it has the property of 'overridingness'[4], but even if this is true at the normative level, there is no reason to expect it to be true of actual motivation. When a moral motivational consideration is in conflict with a non-moral one, there is no reason *a priori* to expect that the moral consideration will win out in what the agent actually does.

At the same time, there is some truth in Hare's claim; if we noticed that someone had *no* tendency to act in accordance with his stated moral convictions, we would eventually wonder whether he had the convictions at all. To return to conscience, although we can appreciate that people can act against their conscience when in the grip of *akrasia*, a systematic disregard for the dictates of conscience would be hard to make sense of. Conscience cannot be systematically motivationally inert. Rather, when we notice a persistent discrepancy between someone's stated convictions and his actions, various other explanations suggest themselves.

One such diagnosis is that of straightforward hypocrisy. Hypocrisy itself is a complex and multifaceted phenomenon. Sometimes it is the cynical presentation of a false image of rectitude or high principles, simply for personal gain; an attempt to free-ride on morality, gaining the benefits of morality in terms of trust and reputation while avoiding its costs wherever possible. Other, more complex kinds of hypocrisy involve a degree of self-deception; hypocrites of this variety may believe that they are not hypocrites and that they do live up to their principles, but through self-deception fail to see the conflict between their principles and practice, perhaps through a self-serving misunderstanding what a particular principle or virtue requires of them. Hypocrites of the first kind do not act against their conscience, since their conscience is a pretence in the first place. Admittedly, the story may be harder to unravel in the case of the self-deceived hypocrite.

However, not all cases of discrepancy between a person's stated convictions and her actions need be explained in terms of hypocrisy. Another possibility is that someone's behaviour may reveal that although she is genuinely committed to certain moral values, she is not fully aware of what her real principles and values are.

Imagine a nurse who is (at least apparently) committed to a fairly rigorous deontological ethics that forbids euthanasia and assisted suicide. She accepts, in an abstract way, that there are cases of terminally ill patients who undergo considerable suffering that cannot be relieved by palliative care, but she also judges that any deliberate killing of a person, even at his repeated, competent request and to alleviate great suffering, can never be justified. In her opinion, all deliberate killing of innocent humans is absolutely forbidden, because of the *kind of act* that it is.

Imagine, however, that in spite of this abstract commitment, when she actually encounters a patient in this desperate condition, she finds herself strongly inclined to 'help the patient die' by assisting her to kill herself. She feels guilt at this reaction, but this does not extinguish her sympathy for the patient, which eventually culminates in her doing the deed. Thus one hypothesis is that although the nurse eventually acts against her own conscience, she does so because she is motivated by compassion, which is a genuine virtue. Her conscience, which has a strongly deontological flavour, continues to give her trouble, but the motivation in acting against it was rooted in a real responsiveness to genuine moral values.

Is it so clear, though, that she really does act against her conscience? There are various ways in which we might analyse this case. We might insist that the nurse acts wrongly (however compassionately) because she acts against her convictions, and leave the matter at that. Or we might just say that she changes her mind when actually presented with the suffering patient, so that her conscience no longer opposes killing in these circumstances. Or we might say that she ends up in a state of conflict, pulled in two opposite directions. All these accounts (and perhaps others) are perfectly plausible. But it is worth exploring another suggestion, which leads us to ask what is really involved in having a moral conviction or a conscientious stance.

In particular, we might make use of a distinction between the 'thin' moral notion of an action simply being wrong or forbidden from the 'thick' moral conception of an action as being generous, kind, imaginative or honest. The distinction between thick and thin ethical concepts is familiar, but I suggest it can also be employed as a tool for finding out what people's deepest moral convictions really are. One may retain the idea of an act as being forbidden, but still feel a moral pull towards features of an act – for example, its kindness – that would suggest it is *prima facie* permissible.

Assisted suicide, of course, is controversial, and the point I really wish to make may easily be obscured by arguments about the pros and cons of this act. So let us turn to something less controversial. Jonathan Bennett[5] describes how in Mark Twain's *The Adventures of Huckleberry Finn*, Huck helps a slave to escape from his 'owner', Miss Watson. Huck is motivated by compassion for the slave, yet feels tremendous guilt about helping him escape. He has helped to deprive the owner of her slave, even though she had never done Huck any harm. As quoted by Bennett, Huck reflected:

> Conscience says to me: 'What had poor Miss Watson done to you, that you could see her nigger go off right under your eyes and never say one single word? What did that poor old woman do to you, that you could treat her so mean? ...' I got to feeling so mean and so miserable I most wished I was dead'.[6]

If we interpret this story as an account of how a man does what he sincerely believes to be wrong even though (as judged by most people nowadays) he clearly does the right thing, and if we also accept Aquinas's account of permissible action described above, then Huck is in a moral pickle. He acts wrongly, because he violates the precepts of his own conscience; however, had he obeyed his conscience and turned the slave in, he would also have acted wrongly (even if his upbringing would have mitigated his blameworthiness) since slavery is a moral evil. However, there is another way of looking at the matter, although it may depart from the author's intentions. This is that the sympathy for the slave that impels Huck to help him is somehow more deeply rooted in Huck's character than his official 'conscience'. Sympathy is a real virtue, and one who is moved by it in circumstances when it does not conflict with other virtues shows a good character, and the true nature of his conscience is revealed.

Of course, there are times when sentimentality masquerades as sympathy, and there are perhaps times when other virtues (for example, justice) conflict with sympathy. But this is arguably not the case here. If a person is unable to let go of his feelings of sympathy (and perhaps justice, or other virtues) in spite of what he claims to believe morally to the contrary, then we should, I suggest, say this: that his real moral values are revealed in these deep dispositions that he cannot give up, and which are enlivened in his imagination by the 'thick' concepts of cruelty, kindness, mercifulness, and so on. And if he thinks about it carefully, he may discover that his values are not what he thought they were.

Likewise, in the case of the nurse, it may turn out that the deontological instinct against euthanasia that she initially holds to is in
reality a moral position in Hare's 'inverted commas' sense of morality –
a 'morality' that is really conceived of as a fallible law or code that one
might quite reasonably decide to reject. In other words, although the
nurse thought she had a moral objection to assisting suicide, what she
understood by this, at first without realising it, was that it was a *commonly accepted* morality, or a code accepted by health care professionals, or the law. It was not morality proper, but rather what was
generally considered to be morality. Her real moral position was
revealed in her sympathetic reaction to her patient in extreme distress
who wanted to be helped to die.

None of this implies that assisted suicide is actually justifiable. The
point of the foregoing discussion, rather, has been to make the perhaps
familiar point that one's real moral values, or the dictates of one's conscience, are not always transparent even to oneself. There can be occasions when what appears to be action in defiance of conscience is
actually the opposite – a discovery of one's real conscientious position
aided by the emergence of hitherto buried sympathies.

These observations do nothing to undermine the thought we started
with, which is that it is necessary but not sufficient to act according to
conscience. Nor is there any suggestion that, in general, people should
follow their 'feelings' rather than their 'reason' when resolving a difficult moral problem. That way lies a corrosive sentimentality. The
claims of justice, for example, can conflict with those of sympathy;
Robin Hood may have robbed the rich to give to the poor out of sympathy, but his actions were arguably still unjust and wrong. The main
point I am making is that one can be mistaken about the content of
one's own moral commitments, be they about the demands of sympathy, justice or any other virtue, and confuse genuine morality with
'inverted commas morality'.

Doing what you think is wrong

Back, then, to the problem of why it should be wrong to do what one
considers to be wrong. We have seen that in some apparent cases of
this, one is not doing what one thinks is wrong. One is doing what
one, deep down, thinks is right, while still being mesmerised (to some
extent) by the knowledge that it is considered wrong, and feeling
guilty accordingly. At the same time, there really are cases when people
act against their conscience. People brazenly, or weakly, do things they

sincerely consider wrong. If their conscience is erroneous, what is their moral fault?

It is hard to answer this except in the terms already suggested – that it shows that morality, in general, does not weigh with them as it should, even if on particular occasions morality does not require what they think it does. If this is correct, what light can it shed on the alleged right of conscientious objection, particularly in health care? If a doctor considers abortion to be murder, does he have the right (or should he be granted the right) to refuse any involvement in abortion, regardless of whether his views are correct? This is a vexing question that is by no means farfetched in clinical practice. Conflicts can arise between the requirements of one's professional role, and one's moral convictions as an individual. Should conscientious objection be respected? And if so, is it only conscientious objection concerning certain particular issues that should be respected, or should it be respected across the board?

Harms, wrongs and conscientious refusal

Various situations come to mind in which it seems, on the face of it, quite unreasonable to grant a right of conscientious objection. The admittedly hackneyed example of abortion is one. In the UK there are legal defences for abortion according to the provisions of the Abortion Act 1967, and in the US it is a constitutional right. But suppose a woman seeking an abortion lives in an isolated rural area where the only doctor(s) she has access to is anti-abortion. She may face tremendous difficulties finding another doctor who is willing to refer her for abortion. Is it fair on the woman to allow the doctor's conscience to stand in her way? The position may well be different in large cities where, if a pregnant woman encounters an anti-abortion doctor, she can probably find another who has no objection, but in isolated areas the situation is arguably far more complex.

Again, what of non-medical examples, such as conscientious objection to fighting in wars perceived to be unjust?[7] When you voluntarily join the Army, you undertake to obey orders, which may include an order to fight in a war whose morality you find dubious. Should you fight in such a war, or not? Of course, you may have naively signed up in the conviction that your country would never fight an unjust war, or that a morally wrong order would never be given. But if that is the case, perhaps you should have known better – it is not clear that your initial complete faith in the Army and your country can excuse disobedience should you become disillusioned.

There are, then, some *prima facie* powerful arguments against being given the right of conscientious refusal in various situations. For many people, however, the intuition persists that such a right is extremely important, and that something of great value is sacrificed if that right is not granted. To explore this powerful intuition, it is worth asking whether, or in what way, a person is either *wronged* or *harmed*[8] by being denied the right of conscientious refusal. What exactly is the nature of his complaint, if he is made to do something he finds morally repellent?

We might initially quibble about what is meant by being 'made to do something'. Literal coercion is one thing, acting under duress is another; acting because of inducements or temptation is different still, and acting after being rationally persuaded is another thing again. In a strictly literal sense, 'coerced action' is not action at all, so no moral blame is appropriate. But talk of coercion is loose, and usually refers to significant duress. So if a doctor is threatened with dismissal if he blows the whistle on a genuinely incompetent colleague, he is under duress to stay silent. His options have been limited to two, both unsatisfactory in different ways. Either he blows the whistle in accordance with his conscience, and suffers dismissal and probably vilification and cold-shouldering, or he keeps his mouth shut, with an uneasy conscience. So our question is: what, if anything, is morally going wrong if he is put under duress to remain silent?

The doctor's initial answer to this question is likely to be simple and obvious: what is going wrong is that he is acting wrongly – in omitting to reveal the truth, he makes a wrong choice. And he might well be correct. But unfortunately, this answer misses the point. Clearly, if the doctor's conscience is *right* to require him to blow the whistle, then he ought to do so, and if he doesn't, he acts wrongly. And if, as is plausible, it is wrong to prevent someone from doing what is morally required of him, then whoever prevents him from blowing the whistle (or more accurately, places him under duress not to do so) also acts wrongly. However, as should be abundantly clear, our real question concerns why it might be wrong to stop someone acting according to conscience, whether his conscience guides him rightly or not. Suppose that the doctor with the troubled conscience is, in reality, an incompetent busybody, self-deluded about his talents and virtues, and always snooping on his colleagues in the hope of discovering blunders they have committed. What then?

Clearly, it is sometimes right to stop people doing things that are in fact wrong and/or harmful, however convinced they are that they are in the right. Indeed, this is the whole premise of the criminal law, as well as professional disciplinary codes. There is a distinction between law and morality, which is particularly marked in liberal democracies. But it would be a mistake to say that the law should never 'enforce morality', and indeed it cannot avoid doing so: murder, rape, blackmail, theft, and a thousand things besides are immoral acts that are rightly prohibited by law. That a man's conscience might impel him to do such things – as is not inconceivable – does not justify us in allowing him to do so. Does this mean that the widespread intuition that a person's conscience must be respected is illusory?

That conclusion would be far too quick. However hard to draw the line, we can at least distinguish between preventing someone from 'conscientiously' doing something harmful to others, and requiring him to do something that his conscience forbids. It is the latter that is really the issue here. The familiar abortion case is particularly relevant. If a 'pro-life' doctor[9] faces sanctions if she refuses to have anything to do with abortion requests, then in effect she is being pressurised to act against her cherished beliefs. And in passing, note that even if there are 'conscience clauses' that allow doctors to refuse to make abortion referrals, in the UK they are still obliged to provide information on how the woman can find another doctor with no such objection. This in itself is problematic, for a doctor who says, in effect, 'I don't refer for abortions because they are murderous, but it so happens that Dr X does, so go to him' is surely to some extent complicit in any abortion that subsequently takes place. If you genuinely think that abortion is murder, just as wicked as the murder of an adult, then you should hardly feel comfortable about facilitating the process whereby someone else commits the murder. I shall not go into this issue further here, but it cries out for extensive discussion.

Back, then, to our main question. What wrong and/or harm (if any) is done to someone who is placed under duress to act against her conscience? The answer cannot be that her conscience guides her rightly, since it may not do so. And if it does guide her rightly, then that alone can account for the wrongness of trying to get her to do otherwise – the fact that it is a conscientious position drops out as irrelevant.

It could be that the real reason is more simple: that pressurising someone to act against her conscience causes her distress, and that is a bad thing. Furthermore, if she succumbs to the pressure and does the deed in question, then she is likely to have a guilty conscience, which is itself distressing. This distress at having done the deed is partly the responsibility of whoever cajoled or threatened her into doing it.

That doing something causes distress is a strong *prima facie* reason to avoid doing it. It can indeed be distressing to be forced to choose between betraying one's deepest values and losing some other important good. Again, non-medical examples illustrate this well. Many a lowly employee is asked to lie for her boss, pretending that he is at meetings when he simply doesn't want to speak to the caller, or falsely assuring suppliers that 'the cheque is in the post'. The lowly employee may, of course, not mind doing this. But often, people do mind being asked to do such things – it distresses them, perhaps because they feel manipulated and bullied, or just because they think it is wrong to lie to cover up for other people's incompetence, laziness or dishonesty. The idea, then, it that it is fundamentally the distress caused that makes such situations morally objectionable.

This consideration is weighty, and no doubt is often sufficient to make it wrong to pressurise people to do what they consider immoral. There is no reason to require a doctor to do something he disapproves of, if there are other doctors who will do the thing in question, without making the first doctor complicit.[10] And the distress this would cause, in these circumstances, is enough to make it positively wrong, rather than merely unnecessary. However, the appeal to distress alone does not seem enough to show that such a doctor would be wronged or harmed by being required to act against conscience. Many things are distressing for doctors (or indeed anyone else) yet they are reasonably required to do them. Breaking bad news is distressing, particularly for sensitive doctors. Watching patients suffer or die may be distressing. The account we are trying to offer still needs beefing up.

One way forward is to examine exactly what the distress is about, when one is distressed at being put under duress to do wrong. It is not, I assume, simply that the wrong is done or that a bad consequence ensues. It is, from a first-person point of view, the fact that the act would be *done by me* that is the heart of the matter. This is the heart of the notion of *integrity*, central to much thought about obedience to conscience. One could even claim that certain assaults on people's integrity, which involve making them complicit in things that go completely against all they love and care for, are not only wrong but *evil*. The Nazis were notorious for facing their victims with unimaginably terrible dilemmas, forcing a mother to choose, for example, which of her two children will go to the gas chamber (and if she refuses to choose, both her children will die).[11] In the former Soviet Union, the relatives of executed prisoners would sometimes have to pay for the bullets with which the prisoners were killed.[12] These are extreme examples, and extraordinarily horrible, but they illustrate how making people become intimately involved in the destruction of those they most love, has a uniquely dark and terrible moral quality, for which the word 'evil' is fitting.

It is thoughts like this that explain why we should not make people do what they think wrong: to do so is a fundamental attack on their integrity, even if not usually nearly as bad as the two examples above. We hear talk of people's 'core values', which define their 'innermost being' and are central to their self-image and self-understanding. We hear how making people act against these values assaults their 'identity'. This language has a tendency to be vague and portentous, but we can grasp the general idea that to pressurise or force someone to involve herself in what she deems morally terrible, is a special kind of attack on her.

Nevertheless, our question has still not quite been answered. For a start, there is a *prima facie* distinction between attacking a person's integrity, and merely attacking her *sense* of integrity. Someone's sense of integrity may depend on things that sensible people regard as trivial, misguided, or even bad. Is it not possible, even, to restore a person's objective integrity by attacking their misguided sense of integrity? We have already agreed that it is right to prevent a person whose conscience requires him to do wicked things from acting on his conscience. Why not go further, and say that it can be right to require a person with a badly misguided conscience positively to act against his conscience? That way we might even get him to see the error of his ways, and get his conscientious position to change. And there is surely no intrinsic objection to trying to get people to alter their moral positions; indeed, it may sometimes be a duty. Most people who hold that it is wrong to get someone to act against his conscientious position, do not believe that it is wrong to employ rational dialogue in order to get someone to alter that position. If that is correct, can we not go further and say that if we can get people to love what is truly right by requiring them, against their will at first, to do what is right, then we are justified in doing so?

In effect, we are still circling around the same question we began with. But although we noted the distinction between one's integrity and one's sense of integrity, we should also note their subtle connections. The basic reason given for saying that it is wrong to do what one considers wrong was that one who does something he believes to be wrong shows a willingness to set aside moral considerations, at least in some circumstances. And if a man does something he thinks is (seriously) wrong, then he will lose some of his sense of integrity. But then, that loss of a sense of integrity is a sign that some genuine integrity has been lost. This is not to say that there is any necessary connection between the two – someone in a severe depression with near-delusional convictions of worthlessness and guilt may retain integrity while losing his sense of it. However, in normal cases there is a connection between the two, simply because it is wrong to do what you consider wrong.

In answer to our guiding question, that of how someone is harmed or wronged by being required to do something against his conscience (and, we might add, acting under that pressure) the best answer, admittedly imprecise, is that such requirements assault his integrity. However, we have to admit that we have not fully analysed integrity and that many more questions are raised. We also face the problem, discussed by Bennett[13] alongside his discussion of Huckleberry Finn, of what to say about someone with a really evil conscience.

In regard to this, Bennett discusses the case of Heinrich Himmler, head of the Nazi SS, whose moral code dictated ruthless cruelty and mass murder on an astonishing scale. Although Bennett insists that the code of conduct Himmler followed, however evil, was indeed a *moral* code, we should note the difference between his code and that of most others whose conscientious positions we disagree with. For example, one difference between Himmler and Huck, who felt guilty about helping the slave escape, is that, as I have suggested, Huck's action revealed deep human sympathies that showed what his moral position really was, even though he was not fully aware of it. With Himmler, as described by Bennett, things were different. Himmler was aware of the violation of sympathies that the extermination of the Jews required, and knew that killing the Jews might corrupt the killers. Yet he also thought that overcoming one's natural sympathies was a mark of Nazi virtue; in this respect if no other, his approach resembles one often attributed to Kant, who thought that the moral worth of an action was most apparent when it went against natural sympathy. Thus although Himmler understood the sympathies that made killing difficult, his conscience clearly directed him to commit and order these cruelties. In a perverse way, what decent people regard as valuable sympathies, Himmler saw as temptations to be resisted. One could experience the temptation, but must still overcome it.

A further difference between Himmler and most others whose moral-ity we disagree with is this. If we disagree, say, with a nurse's stance on euthanasia, we can usually see an intelligible connection between that stance and certain genuine values that we can share. If we oppose euthanasia, we can at least understand that someone who supports it does so because she cares about ending human suffering, and we can agree that in general this is a good thing. If we support euthanasia, we can understand that someone who opposes it does so because she values human life and is repelled by taking it, and can agree that these are good dispositions to have, even if misapplied in this particular case. The same can be said about a variety of conscientious positions with which we disagree, or even find strange. We might think that the stance of Jehovah's Witnesses on blood transfusions is quite irrational, but still see that the steadfastness with which it is held, sometimes at great personal cost, is intelligibly connected with certain religious values of obedience to God that we might (at least if religious ourselves) share. But in the case of Himmler, it seems impossible to connect his moral code with anything we can properly value. There is no connection between the extermination of whole races of people, and any genuine moral value; the Nazi moral code appears to be entirely evil – in intention, execution and consequence.

It is this feature of evil codes and the conscientious following of them (for let us assume that Himmler was not without conscience, but rather had an evil conscience) that stretch the idea we have been defending, which is that people should never be made to act against their conscience. Should Himmler not be made to do a good turn to a Jew, for example? If we cannot give a definitive answer to this, we must stress that it is a limiting case. Respect for the conscience of those we disagree with is more often founded in the knowledge that the people we disagree with do care about morality, and – crucially – have some sense of some basic, genuine moral values. In the normal case, then, we can stick by our claim that it is wrong to act against conscience, and wrong (at least in part, because an attack on integrity) to put people under pressure to act in ways they consider (seriously) wrong. In fact, the position can be stated formally as follows:

1. It is wrong to act in ways that one considers to be wrong.
2. It is wrong to try to get someone to do what is wrong.

From which we can derive:

3. It is wrong to try to get someone to do something that he considers wrong.

This is logically valid, and the argument, in fact, does not explicitly appeal to the idea of harming someone by violating his integrity. What might be controversial is that one wrongs someone by persuading him to do something that is wrong. Those inclined to a Platonic view of this matter might say yes – wrongdoing harms the soul. But even without this supposition, the above argument is free-standing.

Conscience and the health care context

If the above is on the right lines, what should we say about the right of conscientious refusal in a health care context? We have seen some *prima facie* reasons for not granting this right – for example, in the case of the woman who cannot get an abortion because the only doctor she can find objects to it. But at the same time, the argument about the need to respect conscience is powerful. We appear to end up with a simple conflict. Quite apart from the intrinsic morality of abortion, there are weighty reasons both for, and against, making the woman's doctor refer for abortion, assuming this would accord with the relevant legal provisions.

It is very important to note, though, that even if no one should act against their conscience (or be required to), there appears to be no general right to be admitted to the medical or nursing profession. Just as the Army might turn down a candidate who, during the early selection process, turned out to be a pacifist and was applying to the Army either as a practical joke, or because he was labouring under certain fundamental misconceptions about what the Army is, so the medical profession might reject a candidate who said she would refuse to do something normally required by the job. And again, we are not discussing the rights and wrongs of what the profession requires of its members – the issue here is whether recruiters should have *the right* to turn down a candidate who declares a certain ethical position at odds with normal practice.

But the issue is not completely straightforward. We may be able to distinguish between different cases of conscientious objection, and suggest that some conscientious objectors should be admitted to the profession and granted the right to refuse to perform certain procedures, whilst other conscientious objectors should not have that right, and hence should be offered the choice either to be admitted to the profession and comply with its practices, or be refused admission. How might this be done?

Imagine two cases of conscientious objection. One objector is a doctor who is 'pro-life' and will not involve himself in abortion. Another objector, however, is an aspiring obstetrician who turns out to oppose the use of epidurals or any other pain relief during labour. When asked to explain himself, he answers that in the Book of Genesis, it is made clear that the pain of childbirth is a punishment for the sin of Eve in the Garden of Eden. Doctors therefore have no business alleviating this pain. His objection is openly religious (unlike the 'pro-life' doctor who, although a Roman Catholic, plays this down and only uses only secular arguments against abortion) and if refused admission he will certainly have things to say to the Equal Opportunities Commission about the matter, regarding discrimination on the basis of creed.

In discussion of this sort of problem, Mark Wicclair has identified a plausible solution.[14] We have already seen, when discussing Bennett's account of Himmler, that some conscience-driven actions connect in no intelligible way with proper moral values, whereas many others do, even if we disagree with them. A related thought is available here. Medicine has certain core values and practices. There is room for dispute about them and their application; indeed, the discipline of medical ethics would not exist without such arguments. But at the normative epicentre of health care, so to speak, there is concern to alleviate suffering, respect for human life and autonomy, desire to cure disease. Some authors would prefer to describe this ethical heart of health care practice in terms of the Four Principles of respect for autonomy, non-maleficence, beneficence and justice.[15] It is notorious that appealing to these Principles leaves many ethical issues undecided, but whatever their proper application they do seem to define a relatively undisputed set of values. The question of whether to allow conscientious objection may well turn on whether the ethical position of the doctor or nurse connects intelligibly with the core values of medicine.

Whatever our views about abortion, women's rights and the metaphysical status of the embryo or foetus, we can see that the anti-abortion position does connect intelligibly with the core values of medicine. It bases itself on non-maleficence, and regards all killing, including of the unborn, as at least *prima facie* maleficent. It sees itself as an extension of the principle of respect for human life, and whatever our views about the proper scope of moral concern, it is perverse to deny that the foetus is human and is alive. It is a position that is both controversial and has able defenders. In addition, we should also take some account of what has been called the 'fact of reasonable pluralism' – that when well-informed and well-intentioned people disagree about an issue like this, laws and institutions should not take extreme stances. Not only that, but professions thrive when such reasonable pluralism is respected. It is good for the medical profession to have doctors with differing ethical standpoints, to prevent stagnation and stimulate debate. There is good reason, then, to allow anti-abortion doctors into areas of medicine where abortion might be an issue (e.g. General Practice and Obstetrics and Gynaecology), and to grant them the right of conscientious objection.

In contrast to this, it is much harder to see how the values of the anti-epidural doctor connect with the basic values of medicine. No doubt he will argue about what the values of medicine ought to be, and he may do so intelligently. His religious views may even turn out to be right. But the established practice of medicine simply does not accommodate such views, and it is not clear why that established practice should bear the burden of justification, rather than its opponent. Even the appeal to reasonable pluralism stresses the concept of the *reasonable*, and it is inevitable that there will be disputes about which disagreements involve reasonable adversaries, and which do not. Such disputes will not be solved to everyone's satisfaction, but that does not show that the appeal is without value.

In conclusion, there can be little doubt that the notion of conscience should have a central role in medical ethics and ethics generally, and that it has probably not received the attention it deserves. Conscience is not an infallible or oracular source of knowledge, but rather the capacity to reflect on moral questions, especially as they affect one's own conduct. Moreover, although throughout this discussion I have talked about rightly guided and wrongly guided conscience, I have assumed no meta-ethical position about the ontological status of values, concerning their objectivity, cognitivism, or the issue of moral realism. Discussion of these things would have taken me from the matter at hand. At the normative level, however, it is reasonable to say that following one's conscience is, at least in normal circumstances, necessary but not sufficient for acting permissibly, though we should be aware that one's true values are not always transparent to one. When applied to medicine, there is a difficult balance to strike between respect for conscience and the possible risk this poses for patient care, but provided that conscientious objectors are at least broadly in tune with the core values of medicine, or most of them, respect for their objections both preserves integrity and allows an atmosphere of pluralism and debate to flourish within the health care professions.

Notes

1 Aquinas's view on this is well summed up in Anthony Kenny (1987) 'The Conscience of Sir Thomas More', in Anthony Kenny, *The Heritage of Wisdom: Essays in the History of Philosophy* (Oxford: Blackwell).

2 Kenny, *The Heritage of Wisdom*, p. 110.

3 R. M. Hare (1963) *Freedom and Reason* (Oxford: Oxford University Press), ch. 5.

4 R. M. Hare (1981) *Moral Thinking* (Oxford: Oxford University Press), pp. 53ff.

5 Jonathan Bennett (1994) 'The Conscience of Huckleberry Finn', in Peter Singer (ed.) *Ethics* (Oxford: Oxford University Press), pp. 294–305.

6 Bennett, 'The Conscience of Huckleberry Finn', p. 297.

7 A soldier once told me in casual conversation that most members of the British Army were opposed to the venture in Iraq. Yet they still fought there.

8 I do not use the terms 'to wrong' and 'to harm' to mean the same thing. It is possible to wrong someone without harming them (e.g. tell them a benevolent lie) and also to harm someone without wronging them (e.g. inflict a just punishment).

9 I dislike the terms 'pro-life' and 'pro-choice' to describe positions concerning abortion, but use them because they are already so much in circulation. To describe the anti-abortion stance as 'pro-life' misleadingly insinuates that people who are not opposed to abortion are somehow against life, and to describe the view that abortion is not always immoral as 'pro-choice' ignores the fact that the important thing is not to make a choice (which may be inevitable) but rather to make a good choice.

10 I do not offer an analysis of complicity here. In particular, I assume without argument that the requirement not to be actively involved in (perceived) wrongdoing by others does not entail a requirement actively to prevent the wrongdoing by others – e.g. by picketing the operating theatre where abortions are carried out. Clearly, some philosophers will deny that this sort of act-omission distinction with respect to complicity carries any moral weight.

11 This occurs in William Styron's novel *Sophie's Choice*, later made into a film.

12 I owe this example to Eve Garrard.

13 Bennett, 'The Conscience of Huckleberry Finn'.

14 Mark R. Wicclair (2000) 'Conscientious Objection in Medicine', *Bioethics*, 14(3) (July), pp. 205–27.

15 See in particular Raanan Gillon (1985) *Philosophical Medical Ethics* (New York: John Wiley & Sons).

8
The Treatment That Leaves Something to Luck[1]

Nafsika Athanassoulis

Decisions at the end of life are particularly difficult and a huge number of factors can go towards justifying or unjustifying a particular course of action. In this chapter I want to focus in on what is becoming a legally accepted practice and ask whether a distinction which seems to underlie the legal principle can in fact be supported on good philosophical grounds. The distinction relates to allowing incompetent patients to die and the difference between such omissions and directly killing similar patients. The discussion also has implications for the related issue of respecting the right of competent patients to refuse treatment/have treatment withdrawn and how respect for this right compares with cases where competent patients request assistance in dying.

Making decisions on the patient's best interests

Consider the following real case: An infant, let's call her X,[2] suffering from Down's syndrome has a further complication. She has a potentially fatal bowel obstruction, which however can be treated with a fairly straightforward surgical operation. Her parents decided to decline the surgery on the grounds that given her disabilities the kindest thing to do would be to allow her to die. The child's doctors were uneasy with this decision and took the matter to court. The first court's decision was to respect the refusal of treatment, but the Court of Appeal reversed this decision. The Court of Appeal argued that it is a mistake to place so much emphasis on the wishes of the parents as this is fundamentally a decision about what is in the child's best interests. The Court saw itself as having to decide 'whether the life of this child is demonstrably going to be so awful that in effect the child must be condemned to die or whether the life of this child is still so imponderable that it would be wrong for her to be condemned to die',[3] and decided that in this case treatment should go ahead to ensure the child lives.

Of course, one could argue here that the Court made the wrong judgement when considering the awfulness of this child's life, but in a sense this is beside the point. The important conclusion of this case is that it gives guidance on the kind of reasoning one should follow in such cases. In such situations there is a balancing exercise to be carried out: one has to decide whether this patient's life is so terrible it is better for him/her to be allowed to die, or whether the quality of this life is unknown and one should err on the side of caution. Although in the case of infant X the balance came out in favour of treatment, the alternative possibility was left open by the Court when it stated: 'There may be cases ... of severe proved damage where the future is so certain and where the life of the child is so bound to be full of pain and suffering that the court might be driven to a different conclusion'.[4]

Given that this is the kind of approach one should be taking in deciding these cases, it was then only a matter of time before a case would come to the attention of the Courts which merited a different conclusion on the question of the patient's quality of life. In the case of Baby J,[5] a brain-damaged child who was not, however, dying, the Court of Appeal decided that if she were to go into respiratory failure, treatment should be withheld. The Court clarified the following considerations which must be taken into account in deciding such cases:

1. One must start with a presumption in favour of treatment.
2. One must consider the prognosis in terms of pain, suffering and in general quality of life.
3. Decisions should be co-operative between doctors and parents, and made in the best interests of the child.
4. Decisions relate to death as a side-effect, rather than terminating life.

I would like to argue that in some cases consideration 2, along with judgements about what is in the child's best interest as outlined in 3, defeat consideration 1, at which time it then becomes inconsistent to also hold consideration 4. That is, I am going to argue that, under specific circumstances, allowing a patient to die by withholding treatment is, at least, morally equivalent to killing that patient. It may even be the case that killing the patient has some advantages over allowing her to die.

The examples

Charlie is, unluckily, born with severe mental and physical disabilities. Imagine here a most extreme case of disability, which is, however, compatible with life. His life will be hampered by considerable pain and suffering, physical limitations and cognitive disabilities, with related problems such as an inability to become self-aware, inability to communicate and inability to exercise his autonomy or make decisions. However, difficult as his life is going to be, his condition is stable. What this means is that Charlie's life is not under threat from his condition and he does not, at present, require any treatment to sustain life.

However, there is a further complication in that in the first few weeks of life Charlie contracts pneumonia. In general, let us suppose that pneumonia is a fairly easily treatable condition with a good prognosis in terms of returning the patient to his previous state of health and that this treatment is easily available without any adverse cost implications. Charlie's doctors and parents now have to make a decision: should they treat him for his pneumonia or should they allow him to die by not treating him?

Let's start with a presumption in favour of treatment in line with the first consideration above. We then need to make a judgement about Charlie's prognosis in terms of pain, suffering and quality of life. It seems that given the specifics of Charlie's prognosis, and if we accept a general thought that death is not always the worst thing that can happen to a person (and maybe even not a bad thing at all in some circumstances), we should conclude that Charlie's life is of such poor quality that it would be in his best interests to be allowed to die. His quality of life is so poor as to override the presumption for treatment assumed in the first consideration.

Before we go any further, we need to consider two possible objections to this analysis, the first of which is devastating to the argument, while the second can be overcome. One could argue here that the conclusion regarding Charlie's quality of life is arrived at far too quickly. After all life itself is precious and should (always) be preserved. To do this one could rely on a version of the doctrine of the sanctity of life. Since Charlie's life can be preserved fairly easily and straight-forwardly we have a clear obligation to preserve it.

From the point of view of this chapter, little can be said against such a line of argument. Of course, there are excellent discussions on the validity of the doctrine of the sanctity of life itself, but it is not my aim here to go into such arguments. Rather, I am content to accept that anyone who holds to the doctrine of the sanctity of life will fail to be convinced by anything I have to say next as they would reject the very first step in my argument. However, I should point out what is needed in order to reject this first step. A proponent of the doctrine of the sanctity of life would need to reject entirely the idea that the value of life is a commensurable good which can be set against other goods relating to that person's welfare or other considerations. If life is sacred, it is sacred under all circumstances and one has to reject the very suggestion that we are engaging in a balancing act here as nothing could outweigh the value of life. If the value of life cannot be outweighed or even be the kind of thing that one would set against other considerations in the case of quality of life decisions, then, for the sake of consistency, proponents of the doctrine would not be able to admit to any exceptions, e.g. in the case of self-defence, war or punishment.[6] If life is sacred, then it is unconditionally and always so.

For those who see the value of life as an absolute and incommensurable good, this may well be the end of this discussion. For those who do not, there may be a further objection which needs to be overcome before accepting the conclusion that a judgement about Charlie's quality of life can lead to the assumption that he is better off dead. There is, however, a further objection at this stage; one which may be successfully countered. An astute reader, fearing the direction this discussion could take, might want to resist the suggestion that Charlie's case is a case of passive euthanasia, interpreting it instead as a case of futile treatment. I haven't characterised Charlie's case as one of passive euthanasia as yet, but as I will want to do so shortly, I will try to pre-empt this objection by explaining now why we should not interpret this as a case of futile treatment.

Fortunately, there is no need to enter into the complicated debate regarding the most appropriate definition of futility because Charlie's treatment is clearly not futile. The treatment of pneumonia is medically advanced and its effects well documented. Furthermore, the probability that the treatment will have the desired effect, in that it will eradicate the symptoms and return the patient to his previous state of health, is very high. Not only is the treatment's probability of success high, but also the quality of the outcome is good as the treatment does not have any lasting side-effects or long-term repercussions.

A diagnosis of pneumonia then should, in general, have a very good prognosis. However, although the treatment for Charlie's pneumonia is clearly not futile, the intuition that this is a futile case may persist. This is because of a judgement about Charlie's quality of life. I take it that the line of reasoning here might go something as follows: of course the treatment for pneumonia *in general* has a good prognosis, but is it worth putting *this* patient through this treatment given his general state of health? The answer to this question may well be no, given the account of Charlie's state of health in general, but this becomes a judgement about his quality of life. To put it crudely, it is not the treatment for pneumonia which is futile, but rather Charlie's life. Given how terrible Charlie's life is, he would be better off not being treated for the pneumonia and this brings us back where we started: we have passed judgement on Charlie's quality of life and decided he would be better off dead.

That we are making this kind of quality of life judgement for Charlie may become clearer if we look at Britney. Britney suffers from exactly the same infection as Charlie. The treatment for pneumonia would have the same probability of success for Britney as it did for Charlie and the same quality of outcome in terms of returning her to her previous state of health. Britney's case is, however, different from Charlie's in that Britney was born mentally and physically healthy.

In Britney's case one would imagine that the treatment would only be delayed by the time it would take to communicate to her parents the facts about this treatment and ensure they had given valid consent. To even suggest that, similarly to Charlie's case, there is an option of not treating Britney, an option which needs to be considered and perhaps discarded would be very peculiar to say the least. This is because we have easily and unproblematically made a judgement that Britney's life is worth living and we can add to this the thought that it is the role of doctors to provide this treatment. The judgement that Britney's life is worth living and it would be in her best interests to receive treatment to save her life is, I assume, uncontroversial.[7]

Now Britney suffered from the same condition as Charlie, but was treated where he was not. The third case is Douglas. Douglas is like Charlie in that he has similar mental and physical disabilities, affecting his life in terms of pain, suffering and loss of function in similar ways to Charlie's. However, Douglas does not contract pneumonia and the question of treating him or not, never arises.

For the sake of symmetry here one might want to add the case of Anastasia. She's born healthy and suffers no complications, so similarly to Douglas, the question of whether to treat her or not, never arises. So, here are all four cases together:

	Anastasia	**Britney**	**Charlie**	**Douglas**
Initial state of health	Mentally and physically healthy	Mentally and physically healthy	Severely mentally and physically disabled	Severely mentally and physically disabled
Complications	No	Pneumonia	Pneumonia	No
Should we treat?	Question never arises	Yes	No	Question never arises
Outcome	Lives	Lives	Dies	Lives

If the decision not to treat Charlie is based on a judgement about his poor quality of life – that this is so, is apparent from looking at the case of Britney where the question never arises because her quality of life is so good – and Douglas has the same poor quality of life as Charlie, why treat Charlie and Douglas differently in that Charlie dies and Douglas lives?

The argument from the inappropriateness of luck:

Some outcomes, circumstances, influences, and so on are a matter of luck in that they are outside of our control. In this sense winning the lottery is a matter of luck. There is a chain of causation which leads to my winning the lottery, involving physical laws about gravity and the movement of objects which control the spinning of the number balls, as well as numerous factors which have led me to select specific numbers, but this chain is entirely out of my control. I cannot predict it or even become aware of all the factors influencing it.

There are some spheres of human activity where the influence of luck is considered appropriate. In this sense it is appropriate to win the lottery due to luck. What we mean by this is that some unpredictable and uncontrollable factors regulate my winning, but it is quite appropriate that I have no control over them. To control the outcome of the lottery, i.e. to overcome the lucky element involved in choosing the winner, is to cheat. Similarly, it is appropriate to leave some innocuous decisions down to luck, exactly because they are innocuous. If I have to decide between reading a book and going to the cinema, in circumstances where there is no obligation to do or refrain from doing either, and my personal preferences appear to be equally matched in both options, it is appropriate to toss a coin to decide. However, the influence of luck is quite inappropriate in other spheres of human activity. One area which we try to keep as free as possible from the influence of luck is health care.

Of course, many aspects of our health are outside of our control. We have very little control over our basic constitution. In this respect, some agents will be unlucky in that they will be born with diseases and disorders, medical conditions, disabilities and susceptibilities. Although we exercise some control over our lifestyles and how these affect our health, a million other factors, entirely outside of our control, will impact on our health. Finally, neither the development of medical knowledge nor the availability of technological and other resources is within our direct control. However, although our health is a matter of luck, the allocation of health care and how health care decisions are made in general, should not be a matter of luck. The phenomenon of 'postcode lottery', a system under which some patients receive treatment, while others with identical conditions do not, simply because of where they live, is deeply objectionable. It is objectionable because it reduces decisions of who should receive health care to matters of luck. One's postcode is entirely irrelevant to an evaluation of one's medical need or even more broadly to general considerations about the just distribution of precious and scarce medical resources. We are deeply uneasy about postcode treatments exactly because the distribution of health care should not be, like a lottery, a matter of luck. If decisions have to be made between patients, at least they should be made on equitable grounds and not based on some irrelevant characteristic which is outside of our control.

The decision over who dies and when and how they die, should not be based on matters of luck. That is, the decision over who dies should not be controlled by considerations about who happens to need treatment which can be withheld. If we assume that there are at least some cases where quality of life is so poor that the patient is better off dead, then this decision should apply consistently to all other similar patients. Charlie and Douglas are similar in all the relevant respects which would lead us to make a judgement about their lives. Charlie's infection gives us an opportunity to act on this judgement about his quality of life. This option of selecting non-treatment is not available to Douglas, but only because he never contracted the infection in the first place.

Although the examples above are of incompetent patients, similar arguments can be constructed for competent patients. The real cases of Miss B and Diane Pretty may be two such examples. Miss B was a 43-year-old woman paralysed from the neck down as the result of a blood clot lodged in her spinal cord. Although she had only a 1 per cent chance of recovery from paralysis, she was not in pain and was expected to have an otherwise normal life span. Miss B appealed to the British Courts to allow her breathing machine to be switched off, against the wishes of her doctors. Miss B was mentally competent, and did not wish to switch off the machine herself as she thought this would look like suicide and would affect her relatives. The Courts ruled in favour of Miss B, who died shortly after her ventilator was switched off. The Courts ruled that the only question they were entitled to settle was that of the patient's competence. Once a patient has been declared competent, she has the right to refuse treatment, even life-sustaining treatment, even when this goes against the wishes and judgement of her health care team.

However, it is important to note here that Miss B was given all these choices because she happened to be dependent on a ventilator for her continued treatment. Compare Miss B with the case of Diane Pretty. Diane Pretty, 43 years old, terminally ill with motor neuron disease, went to the European Court of Human Rights to seek assurances that her husband would not be prosecuted for helping her to die, an act which she was physically incapable of carrying out herself. She lost her case and died shortly after. Arguably Dianne Pretty's quality of life was significantly worse than Miss B's. Dianne Pretty was severely disabled, but also in pain, terminally ill, with a very short life expectancy, facing the prospect of an imminent and unpleasant death. Miss B was also severely disabled, but in no pain and facing a normal life span. It was Miss B's reasonable quality of life that led her medical team to refuse to switch off her ventilator. If decisions about euthanasia are based on quality of life and judgements about what kind of life is worse than death, then Diane Pretty had a stronger case that her quality of life was poor than Miss B. However, both cases were decided on an element based on luck. Both cases are premised on the assumption that the individual is the best judge of what is in her best interests, but because of an element out of their control one patient has a wider number and different type of choices available to her than the other. Miss B happened to be connected to a ventilator, which gave her the choice of refusing treatment, whereas Diane Pretty, who was not dependent on any on-going treatment at the time, did not have this choice.

It seems that we do allow some patients choice over the manner and timing of their deaths, but this is dependent on elements outside the patient's control. In general, some patients are physically able to commit suicide unassisted and they have the choice to do so. Some patients happen to be dependent on treatment for their continued existence and they have a choice whether to refuse this treatment or not. However, a third category of patients, those incapable of committing suicide, those who are presently not dependent on any life-sustaining treatment, do not get to choose the manner and timing of their deaths.

This seems to me to be the point explicitly raised by Sue Rodriguez. Sue Rodriguez suffered from amyotropic lateral sclerosis. She had a life expectancy of 2-14 months, during which time she was expected to lose the ability to speak, walk, move and swallow, finally losing the ability to breath unassisted. She petitioned the British Columbian and Canadian Supreme Courts for the right to be assisted to die on the grounds that if suicide is not an offence then assisted suicide should not be either. She argued that to maintain that assisted suicide is an offence when suicide isn't, is to discriminate against those who are physically unable to carry the act out themselves.[8]

In cases of competent patients, some people wish to die, perhaps because they are faced with terminal, debilitating, painful conditions. Of these people, some will have the physical ability and means to commit suicide and others will not, e.g. they may be paralysed or require assistance procuring the means to commit suicide. However, having the physical ability to commit suicide, or being in a situation where one's life is dependent on on-going treatment are entirely a matter of luck. The degree of choice afforded to competent patients is thus dependent of factors entirely outside of their control, but making important decisions in health care based on luck is inappropriate to say the least.

The argument from the demands of justice:

Related to the idea of the inappropriateness of allowing luck to influence decisions in health care, is the idea of justice in health care. Considerations of justice recommend trying to eliminate the influence of luck as much as possible. The discussion of moral luck may be of interest here.

When Bernard Williams introduced the term 'moral luck' to modern moral philosophy, he intended it to be an oxymoron.[9] This is because of the apparent tension between the two terms. Morality is about control, responsibility and the appropriateness of praise and blame, whereas luck is about lack of control and the inappropriateness of praise and blame. Cases of moral luck grate against our sensibilities as we generally only hold people responsible for what is under their control. Similarly, there is a tension between justice and luck. At least one aspect of justice relates to fair, equitable and appropriate decision-making; decisions which are based on relevant facts. We can make sense of justice in this way if we contrast it with discrimination. If just treatment is treating equals equally, then discrimination is treating equals, unequally; i.e. picking on an irrelevant characteristic and making it central to our decision. Rejecting a candidate to medical school because of his skin colour is discriminatory as this aspect of the candidate, i.e. his skin colour, is irrelevant to this decision, i.e. whether he would make full use of the training available and become a competent doctor. In this way, allowing decisions to be made based on factors which are down to luck, results in unfair, inequitable and inappropriate outcomes.

In the cases discussed above, choices relating to the manner and timing of one's death are afforded to some patients, but not others, based on factors which are down to luck. The crucial difference between Charlie and Douglas is that the former happened to suffer from an infection whereas the latter did not. The crucial difference between Miss B and Diane Pretty was that the former happened to be dependent on a ventilator. However, both these circumstances were down to luck and neither is significant in determining who should have a choice over their own death. By allowing some patients this kind of choice and denying it to others, by basing the availability of different options on matters of luck, we are discriminating against some patients.

Now, one may wish to reply here that this is a misrepresentation of what motivates the decision to offer choice to Miss B but not to Dianne Pretty. One could argue that the decision is based on the nature of the choice. Miss B's choice is to be allowed to die, whereas Dianne Pretty's request is to be killed. The decision in Charlie's case is to allow him to die, to give Douglas the option of the same outcome his doctors would have to kill him. This concern over the purported difference between killing and letting die seems to underlie the fourth consideration presented by the Courts. In responding to this objection I will need to say quite a bit about the role of intentions, the distinction between acts and omissions and the difference between killing and letting die. This is an enormously complicated topic on which much of value has already been said. One thing that strikes me about the literature on this topic is that very often there is an attempt to generalise about the importance of the presence or absence of particular factors across all cases. Since there are many different and complex cases as soon as theories are postulated counterexamples are usually raised against them. In what follows I will draw heavily on the particulars of the cases above in order to make my point, as these particulars form part of my argument for drawing a distinction here. This does not mean that generalisations are impossible, but rather that in areas where examples are particularly complex we have to be weary of generalisations which are based on abstracting from particulars. As a result the claims I will make, if substantiated, have to be taken to be relevant for these cases, described by reference to these particulars.

Acts, omissions and intentions

Some of the discussion making a case against euthanasia but allowing the withdrawing/withholding of treatment seems to take place in a framework which assumes that the latter practice is somehow innocuous whereas the former is not. This may well be because bodies setting public policy find it easier to have their recommendations accepted if they are crouched in less controversial terms so they tend to avoid the term 'euthanasia' altogether, but there also seems to me to be a more serious move being attempted here. To avoid this trap, I want to start off by defining euthanasia in terms of its motives. The definition of euthanasia must be related to the idea that death is not always the worst thing that could happen to someone and maybe not even be a bad thing in the first place. To make sense of this we have to allow that there are some lives so overwhelmed by pain and suffering that death is a better option and in some cases even a welcome one. Euthanasia then is the practice of bringing about death when this is understood as being good for that person. In this sense, euthanasia is contrasted with murder, as definitionally the motives behind the former are benevolent whereas the motives behind the latter are malevolent. It is these benevolent motives that allow us to make sense of Singer's distinction of involuntary euthanasia.[10] Although it is difficult to think of circumstances which would justify involuntary euthanasia, conceptually this is a different practice from murder, an act which also goes against the person's (expressed or presumed) wish to live but for malevolent motives.

Euthanasia also differs from accidental acts, or acts performed in ignorance, mistakenly, negligently or by those who are incompetent.

In cases of euthanasia the death is not brought about by mistake or in ignorance, but voluntarily and purposefully and is therefore an expression of agency. This means that regardless of whether one thinks that the infamous Harold Shipman was mentally ill or malevolently murderous, he certainly wasn't practising euthanasia on his patients as the British press was fond of claiming during the period when the discoveries relating to the gruesome case were being made.

These distinctions are important as we should avoid prejudging the issue by assuming that examples of killing necessarily involve malevolent motives, whereas letting others die is always carried out from benevolent motives. I take it this is partly the point of Rachels' famous Smith and Jones examples. Both Smith and Jones are morally reprehensible because of their malevolent intentions towards their nephews. These malevolent intentions were shared both by the uncle who killed the child and the one who stood by and watched while he drowned.[11] So although I haven't said much yet on the distinction between killing and letting die as such and the moral status of either, we can say that malevolent killing and malevolent letting die are morally reprehensible and overall both are more morally reprehensible than either benevolent killing or benevolent letting die. So to allow Britney to die in order that one can inherit is worse than allowing Charlie to die based on a judgement about his poor quality of life, and I think this point can be accepted even by those who think that allowing Charlie to die is still wrong.

Having established a connection between euthanasia and benevolent motives and made a general point that killing and letting die from benevolent motives are, at least, less morally reprehensible than killing and letting die from malevolent motives, we now need to look at the distinction between killing and letting die itself. Let's assume that one's motives are benevolent in both the case of killing and letting die, is there a significant moral different between the two practices such that it would justify us in not treating Charlie but would not justify us in killing Douglas?

One of the background reasons why letting die is presented as innocuous whereas killing is not is that it does not involve the agent actually doing anything, whereas killing involves an act. Now there may well be good psychological reasons for why we find it easier to apportion blame to agents for their actions rather than for their omissions, but I am not sure there are good philosophical reasons for this intuition. It may be that in cases of killing it is easy to assign causal responsibility to the agent who, through his actions, has brought about this death, but this begs the question regarding our responsibility for our omissions.

The concept of an omission needs a bit of explaining. Omissions are quite different from non-actions.[12] There are a variety of non-actions I am not performing at the moment, such as not dancing, not hopping on one leg, etc. The list of non-actions would be quite long, but by contrast we can only make sense of omissions within a context. We can only make sense of an omission in the background of the agent's obligations, patterns of activity, standards of normal behaviour, interpretations of other factors in the particular situation, etc. Omissions are rather specific in the sense that, above, I arbitrarily chose dancing and hoping as examples of the many non-actions I could have listed, but standing by and not saving is the only omission relevant in the description of an agent watching a child drown.

Sometimes the distinction between active and passive euthanasia is presented in a way in which we are encouraged to see omissions as non-actions. Non-actions are not relevant to agency as they are not related to what we expect of agents and therefore what we hold them responsible for not doing. There is no normative reason why I am not dancing or hoping right now. Omissions are different; they are related to agency in that we hold people responsible for their omissions, as this failure to act is understood within a background which warrants this action. By this I don't mean that this requirement to act is never defeasible, but that omissions have to be understood with this requirement in mind. If the requirement for action is not part of our understanding of an omission, then omissions, conceptually are no different from non-actions. Since this requirement for action is part of our understanding of omissions good reasons have to be given to defeat it. Of course, in many cases good reasons will be readily available to defeat this requirement for action and the omission will be perfectly morally acceptable.

By encouraging us to see our omissions as non-actions we are encouraged to see them as outside the causal chain of what happened and outside the sphere of responsibility of the agent. However, in Charlie's case inaction was necessary for the outcome, i.e. Charlie's death. Without this omission the outcome would not have occurred. Part of the confusion here is that omissions leave the status quo unchanged, but this is different from saying that the person who omits is not a moral agent as he has not done anything. Non-actions are not doing anything, omissions are omitting to do something and this has a fundamental impact on the outcome.

This idea of omissions not changing anything is further mistaken. The status quo in Charlie's case is that he will die, but we should be careful not to invest the status quo with any particular moral significance. The status quo merely describes how things are and this is not sufficient for concluding that this is how things should remain and that one has done nothing wrong in allowing things to remain as they are. One way of seeing this is by concentrating on one common account of why letting die is innocuous, which is the argument that one is merely letting nature take its course. Donogan describes agents in situations where they omit to act:

> Should he be deprived of all power of action, the situation, including his bodily and mental states, would change according to the laws of nature. His deeds as an agent are either interventions in that natural process or abstentions from intervention. When he intervenes, he can be described as causing whatever would not have occurred had he abstained; and when he abstains, as allowing to happen whatever would not have happened had he intervened. Hence, from the point of view of action, the situation is conceived as passive, and the agent, *qua* agent, as external to it. He is like a *deus ex machina* whose interventions make a difference to what otherwise would naturally come about without them.[13]

However, this invests the course of nature with extraordinary power in determining our moral obligations. Take the case of Britney's doctors. Britney's doctors are in no sense a *deus ex machina*. They are trained and expected to help prolong lives and relieve suffering. There isn't a presumption that they should allow nature take its course and this is evident from our understanding of what it is to be a doctor and what responsibilities this generates towards one's patients. We expect doctors to intervene, to *not* let nature take its course, and expect them to intervene in a specific manner, i.e. to do what is in the patient's best interests.

So it seems that merely relying on the fact that one has omitted to act is not sufficient from absolving us from responsibility. It is wrong to think that we are not responsible for our omissions as they don't actually involve doing anything. Rather, omissions are distinguished from non-actions as they are understood as requirements to act. This requirement to act is generated by the particular circumstances of the case; in the case of medical treatment there is a requirement for doctors to act in their patients' best interests which usually involves treating. Whether the omission is justified or not will depend on whether the requirement for action is defeasible in this case or not. Furthermore, omissions are necessary for the outcome to take place even though they leave the status quo unaffected. What this means is that we cannot judge an omission to be morally acceptable simply because it is an omission, so letting die cannot be morally permissible in comparison to killing simply because it involves an omission. In cases where others have a strong claim to our assistance, such as the case of a patient and her doctor, failing to act and allowing the patient to die is tantamount to abandoning this duty. So we can be as responsible for our omissions as we are for our actions.

Finally, we can now ask whether killing Douglas is morally equivalent to letting Charlie die. If we can be held as responsible for our omissions as for our actions, how should we judge these two cases? I think the answer to this question has to be another question; a question about our intentions in either case. In killing Douglas the doctor's intention is to bring about his death and this intention is benevolent given the background judgement about Douglas' poor quality of life and the badness of death. It may be that the doctor is mistaken about what is in Douglas's best interests, but at least the judgement is made in good faith and its aim is to act in Douglas's best interests (rather than to profit the doctor, or procure organs, or save the family the financial burden, etc.). In letting Charlie die, the doctor's intention is to bring about his death and this intention is benevolent given the background judgement about Charlie's poor quality of life and the badness of death. Given the similarities between Charlie's and Douglas' quality of life and the moral equivalence of killing and letting die in this case, it is unfair to restrict Douglas's options purely because of an element which is down to luck, i.e. the presence of the pneumonia.

Of course, a final point remains to be made. One could argue here that in Charlie's case there is no intention to bring about the death and this is a morally significant different. This difference makes letting Charlie die morally permissible, whereas killing Douglas is still blameworthy; and this is because in Charlie's case the death is foreseen but not intended. Is this a convincing argument in this case? One suggestion in the literature is that if we want to test whether a death is intended or merely foreseen but not intended we should ask whether we would proceed with the act if we knew the death would not occur.[14] Perhaps we could adapt this question and ask whether the doctor would proceed with the omission of treatment if he knew the death would not occur. If omitting to treat Charlie's pneumonia would not result in his death, but rather would result in a deterioration of his overall condition, it seems quite clear that his doctors should not omit to treat him. His doctors have an obligation to act in his best interests and if this treatment would prevent a deterioration in his condition (or even affect an improvement) they should treat him. The decision to not treat only makes sense if accepted as leading to death, and underlying this argument is an assumption that death is not a bad thing or at least not the worst thing that could happen to Charlie. This seems to indicate that the omission is done with the purpose of bringing about death; i.e. the death was intended and not merely foreseen. The doctors are intending the event of the death and the benefit which will result from this death and it is exactly because of the judgement that benefit will result from the death that the omission to treat intends to bring about this outcome.

One could further object here that Charlie's death is not a guaranteed result of the omission in the way that actively killing him brings about a certain death. One could choose not to treat, fully expecting Charlie to die and find out that against all expectations he lives. I am not sure, however, what the force of the point being made here. Surely, one can intend the death, see it as a benefit to the patient and intend for the patient to have this benefit even if, to adapt a slightly different argument, 'it is not, strictly speaking, within one's power to bring them all about'.[15] For we should be held responsible for what we intend to bring about through our actions or omissions, based on a reasonable judgement of what is likely to come about as a result of what we do or omit to do, regardless of whether the result in fact occurs.[16] Whether Charlie's death occurs or not as a result of the intentional omission to treat him is irrelevant to the moral evaluation of the agent. The fact that his doctors benevolently decided that death is preferable to a life of such suffering goes much further in justifying their omission to treat. If the death does not occur due to unforeseen circumstances, then this is regrettable and to avoid this kind of regret, where the outcomes of one's intentional omissions are hostage to luck, active steps to kill Charlie may be preferable to omissions.

Deciding who not to treat in cases where this decision is likely to lead to the death of the patient is a really difficult decision to make. However, if we are willing to make this decision and we can justify the grounds on which it is made, i.e. a judgement about the patient's best interests we should not shy away from accepting its repercussions. We should accept letting die for what it is, i.e. in these cases equivalent to killing. And we should make sure that the same options are available to all patients so that who dies is not a matter of luck. If luck plays a role not only in who happens to need treatment in the first place, but also in who happens to survive even without this treatment, and if the influence of luck in such cases is inappropriate, we should seek to eliminate its influence. Eliminating the influence of luck will require us to make quality of life judgements openly and fully, accepting that the conclusion they lead to is that some patients are better off dead and we should let them die, assist them in dying or kill them.

Notes

1 I wish I could take the entire credit for this title, but those familiar with the literature on moral luck will recognise its inspiration from Lewis's 'The Punishment That Leaves Something to Chance'.
2 The case, *Re B*, is described in detail in J. K. Mason and R. A. McCall Smith (1999) *Law and Medical Ethics* (London: Reed Elsevier), pp. 371–2.
3 L. J. Templeman, *All England Law Reports*, 1990, 927 at 929, cited in Mason and McCall Smith, *Law and Medical Ethics*, p. 371.
4 Templeman, in Mason and McCall Smith, *Law and Medical Ethics*, p. 372.
5 The case is described in detail in Mason and McCall Smith, *Law and Medical Ethics*, pp. 372–3.
6 I take it that McMahan makes a similar point in his discussion of the doctrine of the sanctity of life, as well as offering convincing arguments for rejecting the doctrine itself, J. McMahan (2002) *The Ethics of Killing* (New York: Oxford University Press), pp. 466–7.
7 I should perhaps add 'in this case', as there may be cases where aspects of the treatment may outweigh the good of prolonging Britney's life, for example her parents may be Jehovah's Witnesses and object to a life-saving treatment involving a blood transfusion. In such cases there are complicated questions relating to what is in a patient's best interests and who should make such decisions on behalf of incompetents. However, this is not such a case as, for my purposes, I don't need further complications and just need a case where it is straightforwardly easy to see what is in Britney's best interests.
8 The case is discussed in J. Thomas and W. Waluchow (2002) *Well and Good* (Ontario: Broadview Press).
9 B. Williams (1981) *Moral Luck* (Cambridge: Cambridge University Press).
10 P. Singer (1979) *Practical Ethics* (Cambridge: Cambridge University Press).
11 This seems to me to be the force of Kuhse's defence of Rachels when she brings up the example of the bystander who kills the truck driver rather than watch him die slowly in the fire. H. Kuhse (1999) 'Why Killing is not always Worse – and Sometimes Better – than Letting Die', in H. Kuhse and P. Singer, *Bioethics: an Anthology* (Oxford: Blackwell).
12 In what follows and specifically in making claims about the relationship between omissions and agency as well as how omissions should be understood within a context I am arguing for a specific interpretation of omissions, following J. Feinberg (1984) *Harm to Others* (New York: Oxford

University Press); and P. Smith (19984) 'Allowing, Refraining and Failing: the Structure of Omissions', *Philosophical Studies*, 45; and (1990) 'Contemplating Failure', *Philosophical Studies*, 59.

13 A. Donogan (1977) *The Theory of Morality* (Chicago: Chicago University Press), pp. 42–3.

14 F. Kamm (1992) 'Non-Consequentialism, the Person as an End-in-Itself and the Significance of Status', *Philosophy and Public Affairs*, 21(4), p. 377.

15 J. McMahan (1994) 'Revising the Doctrine of Double Effect', *Journal of Applied Philosophy*, 11(2), p. 205.

16 I have argued for this in detail in N. Athanassoulis (2005) *Morality, Moral Luck and Responsibility: Fortune's Web* (Basingstoke: Palgrave Macmillan).

9
Passive Death

Ray Frey

In November 2001, the US Department of Justice, under Attorney General John Ashcroft, issued a directive to the effect that assisting suicide is not a 'legitimate medical purpose'. The sanction that accompanied this directive, so far as doctors were concerned, was that, for any physician who prescribed drugs to assist suicide, an action could be undertaken by government to remove the doctor's right to prescribe drugs, thereby effectively undercutting that doctor's practice of medicine. A court injunction prevented this directive from being implemented, and the Supreme Court, during its 2005 term, will review the lower court's decision. Prescribing pills to assist a patient in ending their life is not a legitimate medical purpose here, even, of course, if the prescribing is done at the voluntary request of a competent patient who wants to end their life.

I have elsewhere argued against the kind of reasoning that underlies this Department of Justice directive and have given the reasons why I support physician-assisted suicide and active voluntary euthanasia.[1] Here, I want to look further at one aspect of this debate, namely, withdrawal cases.

Euthanasia is the intentional ending of life from a benevolent or kind motive. The person typically is gravely and terminally ill. In the contemporary debate over euthanasia, however, it is no longer the case that the person must be in imminent danger of dying. Rather, the gravely ill person simply does not wish to continue to live the kind of life to which disease and illness have condemned them. Examples abound, involving AIDS, amyotrophic lateral sclerosis, various cancers, etc., in which, while still alive, a person does not regard themselves as having a life of a quality that they wish to continue to endure. Accordingly, they request assistance in dying, and, quite naturally, the person of whom this request is usually made is a doctor.

To someone who objects to the introduction of quality of life concerns into a debate about life itself, it is hard to make clear how this voluntary request by a patient has a firm ground. Everyone sympathises with gravely ill people and what they are going through, but the mere fact that the quality of one's life has plummeted, perhaps to a disastrous extent, does not to some provide a basis for seeking to end life itself. To others, of course, quality of life concerns are centrally important to the whole issue of the prolongation of life, where a patient, by disease or illness has now reached a level of quality of life with which he does not want to live. What the omnipresent medical cases have done, in the main, is to make more generally known to the public the level to which some lives can plummet and, accordingly, the general thought that a life of that level is no longer worth living. It used to be objected to utilitarianism that it pronounced on whether someone's life was worth living; in the medical cases featured in the physician-assisted suicide literature, it is the very person living the life in question who pronounces on whether that life is worth living.

Three distinctions dominate contemporary discussion of euthanasia. First, active euthanasia involves taking steps to terminate a life; passive euthanasia involves omitting steps to save a life or withdrawing treatment. Active euthanasia is rejected by many, while passive euthanasia is a common occurrence in our hospitals. Turning off the ventilator of a patient whose further treatment is deemed futile is openly done and is said to amount to a passive death; providing a patient with pills which they then take and end their lives is not openly done and is said to amount to an active death. I have argued that this distinction cannot be sustained. After all, a patient will die if they are injected with a large enough dose of morphine, in the guise of relieving pain, but they will also die if their ventilator is turned off or if their feeding tubes are removed. In the case of a patient whose further treatment is deemed futile, what is the difference between these ways of producing death? Sometimes, it is suggested that in active death the doctor causes death, whereas in passive death the doctor allows or permits death (I have explored this causal claim elsewhere.[2] But in the examples I have given, it is the doctor who turns off the ventilator and withdraws feeding tubes. These things do not occur by accident, mistake, ignorance, or negligence; they are knowingly, deliberately done by the doctor. To say in these cases that the patient's underlying illness kills them is simply false; what kills them in the morphine case is respiratory depression, in the ventilator case suffocation, and in the feeding tubes case starvation.

Second, a distinction can be drawn among voluntary, involuntary, and non-voluntary forms of euthanasia. In cases of voluntary euthanasia, the steps taken to end life are taken with the consent of the patient and often, as above, at the request of the patient. In the United States, living wills or advanced directives often convey this consent and may have legal force. In cases of involuntary euthanasia, the steps taken to end life are not taken with the consent of the patient, are not taken in response to requests by the patient; indeed, these steps may well be something that the patient would utterly reject and never endorse. In cases of non-voluntary euthanasia, the steps taken to end life are taken with regard of the life of someone whom medical (and legal) authorities deem incompetent. This person is represented by an authorised trustee, and it is the trustee who consents to the steps that end life. Non-voluntary euthanasia can quickly run into difficulties, if it turns out that a patient has no authorised trustee and so no one authorised by the patient to consent to what is proposed be done. For it can become unclear, if a trustee is then assigned by the courts and if the trustee consents to end life, whether death has become involuntary.

Third, a distinction can be drawn between physician-assisted suicide and euthanasia. I think the clearest way to draw this distinction is in terms of who acts last, causally. In physician-assisted suicide, the patient is the last causal actor; in euthanasia, the doctor is. Suppose a competent patient voluntarily requests assistance in dying and his doctor supplies him with a pill that, if swallowed, will produce death: if the patient swallows the pill, he produces his own death. The doctor does not kill him. The pill only produces death if it is swallowed, and the patient voluntarily swallows it. The death is physician-assisted, since the doctor has supplied the means of death; but the doctor has not forced the patient to swallow the pill or otherwise coerced the death that ensues. So much of the literature against physician-assisted suicide goes on about doctors killing their patients, when, in fact, such cases do not involve doctors killing their patients at all. The last causal actor is the patient.

What is at issue in the contemporary debates is whether laws against physician-assisted suicide and active voluntary euthanasia should be relaxed, in the light of the ravaged lives that illness and disease have forced upon some people. In this context of relaxation are invoked all kinds of slippery slope arguments. Start with voluntary euthanasia and we will soon reach involuntary euthanasia; start with competent patients requesting death and we will soon reach incompetent patients who do not know what they are requesting; start with doctors prescribing pills for competent patients who voluntarily request assistance in dying and we will soon reach conditions in which doctors begin killing off their patients in all kinds of circumstances; start with making assisted suicide a possible option and we will soon reach conditions in which patients feel pressured to choose death, even when they want to live. We need evidence either that these slopes will be in fact be descended or that the likelihood of descending is very much increased if we adopt measures to relax current prohibitions. In most instances, slippery slope concerns are pressed not only against instituting changes in the law but even beginning to think about changes in the law, and this has been especially true in the context of active voluntary euthanasia. It is hard to have a discussion of this particular topic without the Nazi camps being put forward as the inevitable end of this slope of active killing. Safeguards against going down such a slope of killing are doomed to failure; it is better – and safer – not to budge from a position of absolute prohibition. (Interestingly, it seems unlikely that many of those opposed to physician-assisted suicide and active voluntary euthanasia would remove their opposition, were we to find a series of cases in which slippery slope effects could be shown to be very unlikely to arise. For it is not in the end the prospect of severe slope consequences that in their eyes makes these things wrong; such consequences merely add to the unattractiveness of relaxing legal prohibitions.)

Now one very important component in this discussion of euthanasia is whether a patient is regarded as having a right to refuse treatment. In the United States, individual states grant this legal right, and doctors and others have to respect this right. If one has such a right, and if one remains able to insist upon its observance, then it is the not the case that one need live out the life of massively reduced quality to which illness and disease has condemned one. If one is held to have no such right, or if one's illness has progressed to an extent that one can no longer insist upon its observance, or if one has no advanced directive that indicates one's refusal of treatment in certain conditions (or that directive is challenged in the courts), then, rather strangely, one is forced to live out a life that other people wished lived out though the individual in question does not. If, for the sake of argument, we take these other people to be the doctor, then we have the very odd situation in which a patient must live out a life that his doctor deems worth living out. This not merely raises issues to do with patient autonomy and control over one's own life, but it seems to imply that the doctor has somehow gained moral control over me and my life; I must live out whatever life he deems I should live out, whatever I think about the matter.

Who, it might be asked, objects to patient refusals of treatment? I suspect many deontologists and most religious people do, to the extent that they regard refusal of treatment as equivalent to suicide. For suicide is forbidden. I must continue to live, whether I wish to or not; morally, I do not have the choice of death even by refusing treatment. Thus, even if one were to continue to bar physician-assisted suicide and active voluntary euthanasia, I am not even permitted to end my life through refusing further treatment, when I judge my life to be no longer worth living. To do so is, in effect, to commit suicide, and suicide is forbidden. Of course, to the extent that one can argue that refusing treatment is not suicide, then to that extent perhaps the door of having a say in ending my life may open a bit. Still, every gravely ill, competent patient knows what is almost certainly going to ensue if they refuse further treatment; if they then decide to carry on and refuse life-saving treatment, it is hard to see why they may not be said to intend to die. Indeed, the patient may vigorously protest if the doctor seeks to save them and demand that his refusal of treatment be honoured.

If refusing treatment is, however, suicide, and one may permissibly refuse treatment, then a puzzle arises. If patient suicide is permissible, then what makes suicide by means of the doctor-supplied pill impermissible? One agrees that patients may refuse treatment and die; one agrees that competent patients know they will die, if they refuse life-saving treatment. So why is refusing treatment permissible but taking the pill not? If one may not take one's life, then one may not refuse life-saving treatment; but if one may refuse life-saving treatment, as in the United States one may, then why may not one take a pill that produces death? To respond now that it is not suicide which is forbidden but doctors killing their patients is not to the point; in neither the refusal nor the pill case does the doctor kill his patient. The patient kills the patient. It seems completely arbitrary to permit suicide but to favour one way of ending life over another.

Nor does it seem to the point to argue that what is objectionable is not taking one's life but supplying the means of death to the patient. For why is it wrong to supply the means of death? Surely the answer turns upon the fact that the patient will use that means to take their life. But if taking one's life is permissible, why is it wrong to supply the pill? (I do not have space here to go over the usual arguments given for why suicide is wrong, ranging from claims about life being a gift from

God to claims about deserting one's post, including one's family and friends. I have not found these arguments compelling.)

What is the fear, ultimately, in the doctor's case? It is, as I have sketched it, the fear that he may be a cause of death. The claim upon which a great deal of the euthanasia discussion hinges is that, if he withdraws feeding tubes or turns off a ventilator, he is not a cause of death, whereas if he supplies the pill to the patient who then voluntarily swallows it, he is a cause of death. In passive death, he is not a cause; in active death, he is. And, of course, if he is a cause of death, then he may be held (partially) responsible for a death, morally and perhaps legally. But how are we supposed to construe what he does, in injecting morphine to relieve pain, in turning off a ventilator, and in withdrawing feeding tubes? These are not omissions on his part; these are actions on his part. Nor can we treat the cases as ones involving only the patient, as if the doctor were not present. The doctor *is* present and *is* the one who injects the morphine, turns off the ventilator, and withdraws feeding tubes; death ensues only after he does these things. The patient's autonomous consent to what the doctor does as a result of the patient's refusal of further treatment does not make the autonomous, voluntary decision by the patient to forgo further treatment into the only morally relevant fact in the situation. Legally, in a particular jurisdiction, it may be permissible for the doctor under specified circumstances to withdraw feeding tubes; indeed, in the light of the patient's refusal of further treatment, we may want the doctor to withdraw feeding tubes. But, morally, that in no way shows that the doctor's withdrawal is not partly morally responsible for the patient's death, and it becomes morally relevant in this way because the act of injecting morphine, turning off the ventilator, or withdrawing feeding tubes helps cause the patient's death. This does not confer guilt upon the doctor; it simply affirms that what one causes in the world is relevant to what one is morally responsible for. Indeed, we may want the doctor to take seriously the autonomous, voluntary decision of his patient to refuse further treatment; but the fact that we want the doctor to do this does not settle the issue of whether, say, withdrawing feeding tubes caused death by starvation. Nor does the fact that the law may hold that it is legally permissible for the doctor to withdraw feeding tubes show that the withdrawal of feeding tubes did not cause death. How did starvation come to kill the patient? The doctor withdrew the patient's feeding tubes, admittedly, in the light of the patient's valid refusal of further treatment; but the patient's valid refusal does not relieve the doctor of causation in the affair, only a kind of legal charge for having produced a death.

In the withdrawal case, there is a temptation, in order to try to remove the doctor from a causal role in what befalls the patient, to argue
either that the patient's underlying illness kills him, which is false, or
that the doctor, legally permitted to do what he does, does not help
cause death, which is also false. What requires stress here is not that,
but for the withdrawal of feeding tubes, the patient would not have
died of starvation, though this is true; it is that what enables starvation
to enter the causal picture at all, what permits it to serve a causal role
in the patient's death is the doctor's withdrawal of feeding tubes. What
appeal to law and legality cannot remove is this fact, that starvation is
only able to overtake the patient and cause death if the doctor acts one
way as opposed to the other. This, I think, is the crucial point: starvation only gets into a position to be fatal to the patient through the
doctor's act of withdrawing feeding tubes. This is why it is so misleading to say that it is the patient's underlying illness which kills him: this
way of talking makes it appear as if starvation somehow got into a
position to cause the patient's death without any action on the
doctor's part. This is false. A vital part of what befalls the patient is left
out, much as would be true in a case in which someone fell off a cliff,
without it being added that someone pushed him. The fall just did not
kill this individual; he was put into a position where the fall could kill
him, and how he got into that position, where the fall could produce
his death, is part of the causal story of that individual's death. The
same is true of the doctor's injection of morphine, or turning off the
ventilator, or withdrawing feeding tubes.

Notice how the doctor's intention is not relevant to this discussion.
Suppose two patients die of respiratory depression and two doctors
have injected each of their respective patients with a largish dose of
morphine; suppose further that the intention of one doctor is to
relieve his patient's suffering and the intention of the other is obtain
money through the patient's will; in both cases, the cause of death is
respiratory depression brought on by the largish dose of morphine. It is
true that one will want to say of the one doctor that he not only causes
death but also is guilty of murder, but his intention in injecting the
morphine does not matter to whether the injection of morphine is a
cause of death. And the morphine only gets into a position to be a
cause of death through the doctor's injecting it. Similarly, the doctor
who turns off the ventilator is a cause of death, whatever their intention in turning it off. The patient is dead because of inability to
breathe, which causes death. Turning off the ventilator enables the
patient's difficulty in breathing to kill him; put differently, this diffi-

culty in breathing gets into a position to kill the patient as a result of the doctor's turning off the ventilator. Likewise, with the feeding tubes: the doctor's act of withdrawing them is a cause of death. His intention in withdrawing them may affect other things we may go on to say of his act, but his intention does not go to the issue of whether his act causes death by starvation. So, what of the patient's consent to the withdrawal? It goes towards the issue of whether a moral (and legal) charge should be brought against the doctor for doing to the patient something that the patient refused to have done; it does not go the issue of whether the withdrawal of feeding tubes causes death.

Withdrawal of feeding tubes, then, is not an alternative to physician-assisted suicide, so far as causality is concerned. The doctor takes the essential step in enabling starvation to kill his patient. To say in the pill case that the doctor also takes an essential step in enabling his patient to kill himself does not show any causal difference between the withdrawal and pill cases, or between passive and active death. In this way, nothing really turns upon the death being a passive one.

Nor is anything achieved by trying to carve out another difference in intention. It might be claimed, for example, that in supplying the pill, the doctor must intend the patient's death, whereas in the withdrawal case he may well not intend this. But in the pill case he may well not intend death; he may take himself only to be supplying a means of death to someone who will not take the pill or who wants it only for reassurance, if he continues to deteriorate. And in the withdrawal case, he may well intend death and withdrawal of feeding tubes as the way of achieving this. In both cases, the doctor remains a part cause of a death. Moreover, what those who opposed physician-assisted suicide and active voluntary euthanasia have insisted upon is that doctors may not kill, not that they may not intend death. In the pill case, even if the doctor intends death, the pill can only kill the patient if the patient decides to swallow it. The doctor does not kill the patient, whatever the intention of the doctor.

To those who oppose physician-assisted suicide and active voluntary euthanasia, withdrawal is seen as an alternative, one that is supposed to be grounded in differences in causal structure between withdrawal and, say, pill cases. I do not think there is any difference in causal structure.

Finally, if I want the doctor to pay attention to the patient's refusal of further treatment, and if, for example, I agree to underwrite this concern with a law that cedes the patient a right to refuse treatment, then why may not the doctor argue that he is powerless in the situation? What is at work, so to speak, is the will of patient? He, the doctor, is powerless in the situation; he simply does what the patient instructs him to do. This is again, however, to misconstrue the matter. The may indeed see himself as forced to adhere to the law, but he cannot use this fact to escape causal responsibility for the death that ensues as a result of his deciding to withdraw feeding tubes. He could choose not to go ahead. Unquestionably, this places the doctor in a difficult position: if he ignores his patient's refusal of further treatment, he may run into the law, whereas if he honours his patient's refusal and withdraws feeding tubes, he becomes causally, and so perhaps morally, responsible for his patient's death. This, I think, is the doctor's actual situation.

Interestingly, consent does not separate the withdrawal and pill cases. In the withdrawal case, the patient consents to what the doctor does to him. In the pill case, the patient asks for what the pill the doctor prescribes, chooses then to swallow the pill, and dies as a result. It is not as if the pill is foisted on to the patient or that it produces death without its having had to be swallowed by the patient; and it certainly is not true that the pill is crammed into the patient's mouth.

Thus, I do not see withdrawal as an alternative to physician-assisted suicide and active voluntary euthanasia because I think the cases are alike, causally, without their being any other significant moral differences between them. I do not think anything is gained, therefore, by trying to make out passive death as something quite different from active death, in the context of the withdrawal and pill cases.

One final point. Suppose one argues, as I indicated earlier, that one simply rejects the permissibility of suicide and so rejects the permissibility of refusal of treatment how is one to argue the withdrawal case? If suicide is impermissible, what could possibly make the doctor's withdrawal of feeding tubes permissible? Withdrawal then looks like an attack on the life of the patient. Suppose the doctor decides that further treatment is then futile and withdraws feeding tubes: this seems to make it obvious that the doctor's decision to withdraw is what produces death, without the intervention by anything on the patient's part to relieve the doctor's causal situation. Suppose, then, one argues that the withdrawal case is not one of suicide, that insisting upon a right to refuse treatment is not a way of committing suicide: the problem now is to understand exactly how we are to construe what the patient is doing. To say that the patient is just trying to escape the misery of his circumstances is to leave out how he understands that escape to take place. The patient is competent and knows what refusing food and nutrition means for him. Well, it might be claimed, he does not intend to die? But in this context how are we to understand what he takes himself to be doing in refusing treatment? This is not the situation in which there is a debate about intention and foresight; for what we are trying to understand is what the patient sees his right to refuse treatment to be doing in his case. He sees it as a way of ending his life and he embraces this way. The reason refusing treatment is so important in his case is because he can end his a life that illness and disease has led him to think he no longer wants to live. There does not appear to be any absence of intention on his part, when it comes to ending his life.

Notes

1 Gerald Dworkin, R. G. Frey and Sissela Bok (1998) *Euthanasia and Physician-Assisted Suicide* (Cambridge, Cambridge University Press).
2 See R. G. Frey (2005) 'The Doctrine of Double Effect', in R. G. Frey and Christopher Wellman (eds.), *The Companion to Applied Ethics* (Oxford: Basil Blackwell); and R. G. Frey (forthcoming) 'Intending and Causing', *Theory of Ethics*.

Index